UNRULY PLACES

UNRULY PLACES

LOST SPACES,
SECRET CITIES,
and Other Inscrutable
GEOGRAPHIES

ALASTAIR BONNETT

VIKING

VIKING

an imprint of Penguin Canada Books Inc., a Penguin Random House Company

Published by the Penguin Group
Penguin Canada Books Inc.
90 Eglinton Avenue East, Suite 700, Toronto, Ontario, Canada M4P 2Y3

Penguin Group (USA) LLC, 375 Hudson Street, New York, New York 10014, U.S.A.
Penguin Books Ltd, 80 Strand, London WC2R 0RL, England
Penguin Ireland, 25 St Stephen's Green, Dublin 2, Ireland (a division of Penguin Books Ltd)
Penguin Group (Australia), 707 Collins Street, Melbourne, Victoria 3008, Australia
(a division of Pearson Australia Group Pty Ltd)
Penguin Books India Pvt Ltd, 11 Community Centre, Panchsheel Park,
New Delhi – 110 017, India
Penguin Group (NZ), 67 Apollo Drive, Rosedale, Auckland 0632, New Zealand
(a division of Pearson New Zealand Ltd)
Penguin Books (South Africa) (Pty) Ltd, 24 Sturdee Avenue, Rosebank,
Johannesburg 2196, South Africa

Penguin Books Ltd, Registered Offices: 80 Strand, London WC2R 0RL, England

First published in Great Britain in 2014 by Aurum Press. Published in Viking hardcover by
Penguin Canada Books Inc., 2014. Simultaneously published in the U.S.A. by Houghton Mifflin
Harcourt Publishing Company, 215 Park Avenue South, New York, NY 10003.

1 2 3 4 5 6 7 8 9 10   (RRD)

Book design by Alex Camlin
Manufactured in the U.S.A.

LIBRARY AND ARCHIVES CANADA CATALOGUING IN PUBLICATION

Bonnett, Alastair, 1964–, author
Unruly places : lost spaces, secret cities, and other
inscrutable geographies / Alastair Bonnett.

Includes bibliographical references and index.
ISBN 978-0-670-06718-3 (bound)

1. Geography--Miscellanea. I. Title.

G131.B65 2014   910   C2014-902912-8

eBook ISBN 978-0-14-319206-0

British Library Cataloguing in Publication data available
American Library of Congress Cataloging in Publication data available

Visit the Penguin Canada website at **www.penguin.ca**

Special and corporate bulk purchase rates available; please see
www.penguin.ca/corporatesales or call 1-800-810-3104.

For
Helen
and
Paul

Contents

We are headed for uncharted territory, to places found on few maps and sometimes on none. They are real places but extraordinary ones. For the most part they are unvisited and passed by, but each of them shows that there are places that can still surprise and disorient us in a world where Google Maps seeks to map out every last inch.

There is another Canada, somewhere in ruins. Over recent years, abandoned mining towns, like Pine Point in Northwest Territories, or Newfoundland's "outports" (remote fishing villages that were first emptied out in the decades after Newfoundland joined Canada) have begun to nag the imagination. A romantic aesthetic of loss has been at work, sending photographers off to capture the poignancy of decay. In our melancholic era, it is towards such uncanny places, the losers in the march of progress, that our sympathy turns. Today we don't fear ruins and decay as much as banality, the generic landscapes that surround, sustain, and suffocate us. Perhaps the most truly lost places in Canada are the snips of dirty scrub that lie in between its urban expressways.

I date my interest in the strangeness and ubiquity of the modern world's unruly places to when I lived on the outskirts of Hamilton, Ontario, in the early 1990s. I didn't have a car and chose to walk, often finding myself alone on benighted islands amid the churning traffic. It came as no surprise to me that the hobby of "urban exploration," of voyaging the forbidden zones of the city, arose from within southern Ontario's metropolitan sprawl. Toronto's *Infiltration*—"the zine about going places you're not supposed to go"—was in at the start of what is now a worldwide phenomenon. The city can become a "wonderful playground," declared its founder, the legendary Ninjalicious.

Websites dedicated to the exploration of the hidden places and "subscapes" of urban Canada now proliferate. My favorite is Vanishingpoint, which is dedicated to drain and sewer systems, and includes some revelatory accounts of journeys underneath Niagara Falls. "Imagine a tunnel more than ten storeys underground, a hundred years old, bricklined, wet, and completely inaccessible save by descending through a narrow slit in its ceiling," recalls Vanishingpoint voyager Michael Cook, and "Now imagine that this tunnel flows into Niagara Falls."

The re-enchantment of place feels ever more necessary, but my time in Hamilton impressed on me that many of the places that challenge and intrigue us are not spectacular or even concealed. Most of the sites in this book are in plain sight, many are full of people, but they all have the capacity to change how we think about place. Sometimes something as small as a change of name can be enough, displacing the past but not neutralizing it, interring "Leningrad" under the soil of "St. Petersburg." The weight of the past subtly subverts modern places, nagging at them about something buried. It's obvious to us why the town of Pile o' Bones wanted to change its name, becoming Regina in 1882, but it's a rupture that creates a gnawing absence in the town's collective memory. The old name was of Cree origin and referred to the bison bones piled up on Wascana Creek. A change of name may seem a weightless thing when compared to some of the dead cities and dark labyrinths that we will be roaming through. Yet sometimes the most unruly places are those that are both close by yet just out of reach.

INTRODUCTION

Our fascination with remarkable places is as old as geography. Eratosthenes's *Geographika,* written around 200 B.C., offers a tour of numerous "famous" cities and "great" rivers, while the seventeen volumes of Strabo's *Geography,* written in the first years of the first century A.D. for Roman imperial administrators, provides an exhaustive compendium of journeys, cities, and destinations. My favorite of Strabo's places are the gold mines of India, which, he tells us, are dug by ants "no smaller than foxes" that possess pelts "like those of leopards." Although our appetite for curious tales from afar has been continuous, today our need for geographical reenchantment is of a different order.

I root my love of place in Epping. It's one of many commuter towns near London, pleasant enough but generic and placeless. It's where I was born and grew up. As I used to rattle out to Epping on the Central Line or drive there along London's orbital motorway, I often felt as if I were traveling from nowhere to nowhere. Moving through landscapes that once meant something, perhaps an awful lot, but have been reduced to spaces of transit where everything is temporary and everyone is just passing through, gave me a sense of unease and a hunger for places that matter.

You don't have to walk far into our coagulated roadscape to realize that, over the past one hundred years or so and across the world, we have become much better at destroying places than building them. The titles of a clutch of recent books, such as Paul Kingsnorth's *Real England,* Marc Augé's *Non-Places,* and James Kunstler's *The Geography of Nowhere,* indicate an emergent anxiety. These authors are tapping into a widespread feeling that the replacement of unique and distinct places by generic blandscapes is severing us from something important. One of the world's most eminent thinkers on place, Edward Casey, a professor of philosophy at Stony Brook University, argues that "the encroachment of an indifferent sameness-of-place on a global scale" is eating away at our sense of self and "makes the human subject long for a diversity of places." Casey casts a skeptical eye over the intellectual drift away from thinking about place. In ancient and medieval thought place was often center stage, the ground and context for everything else. Aristotle thought place should "take precedence of all other things" because place gives order to the world. Casey tells us that Aristotle claimed that place "gives bountiful aegis — active protective support — to what it locates." But the universalist pretensions of first, monotheistic religion, and then the Enlightenment, conspired to represent place as parochial, as a prosaic footnote when compared to their grand but abstract visions of global oneness. Most modern intellectuals and scientists have hardly any interest in place, for they consider their theories to be applicable everywhere. Place was demoted and displaced, a process that was helped on its way by the rise of its slightly pompous and suitably abstract geographical rival, the idea of "space." Space sounds modern in a way place doesn't: it evokes mobility and the absence of restrictions; it promises empty landscapes filled with promise. When confronted with the filled-in busyness and oddity of place, the reaction of modern societies has been to straighten and rationalize, to prioritize connections and erase obstacles, to overcome place with space.

In his philosophical history *The Fate of Place* Casey charts a growing "disdain for the genus loci: indifference to the specialness of place." We all live with the results. Most of us can see them outside the window. In a hypermobile world, a love of place can easily be cast as passé, even reactionary. When human fulfillment is measured out in air miles and when even geographers subscribe to the idea, as expressed by Professor William J. Mitchell of MIT, that "communities increasingly find their common ground in cyberspace rather than *terra firma*," wanting to think about place can seem a little perverse. Yet placelessness is neither intellectually nor emotionally satisfying. Sir Thomas More's Greek neologism *utopia* may translate as "no place," but a placeless world is a dystopian prospect.

Place is a protean and fundamental aspect of what it is to be human. We are a place-making and place-loving species. The renowned evolutionary biologist Edward O. Wilson talks about the innate and biologically necessary human love of living things as "biophilia." He suggests that biophilia both connects us together as a species and bonds us to the rest of nature. I would argue that there is an unjustly ignored and equally important geographical equivalent, "topophilia," or love of place. The word was coined by the Chinese-American geographer Yi-Fu Tuan about the same time as Wilson introduced biophilia, and its pursuit is at the heart of this book.

There is another theme that threads its way throughout the places corralled here—the need to escape. This urge is more widespread today than at any point in the past: since fantastic vacation destinations and lifestyles are constantly dangled before us, it's no surprise so many feel dissatisfied with their daily routine. The rise of placelessness, on top of the sense that the whole planet is now minutely known and surveilled, has given this dissatisfaction a radical edge, creating an appetite to find places that are off the map and that are somehow secret, or at least have the power to surprise us.

When describing the village of Ishmael's native ally and friend, Queequeg, in *Moby-Dick,* Herman Melville wrote, "It is not down in any map; true places never are." It's an odd thing to say, but I think it makes immediate, instinctual sense. It touches on a suspicion that lies just beneath the rational surface of civilization. When the world has been fully codified and collated, when ambivalences and ambiguities have been so sponged away that we know exactly and objectively where everything is and what it is called, a sense of loss arises. The claim to completeness causes us to mourn the possibility of exploration and muse endlessly on the hope of novelty and escape. It is within this context that the unnamed and discarded places—both far away and those that we pass by every day—take on a romantic aura. In a fully discovered world exploration does not stop; it just has to be reinvented.

In the early 1990s I got involved with one of the more outré forms of this reinvention, known as psychogeography. Most of the time this involved either drifting in search of what some of my comrades fondly imagined were occult energies or purposely getting lost by using a map of one place to navigate oneself around another. To wander through a day care center in Newcastle while clutching a map of the Berlin subway is genuinely disorienting. In so doing, we thought we were terribly bold, but in hindsight what strikes me about the yearning to radically rediscover the landscape around us is just how ordinary it is. The need for reenchantment is something we all share.

So let's go on a journey—to the ends of the earth and the other side of the street, as far as we need to go to get away from the familiar and the routine. Good or bad, scary or wonderful, we need unruly places that defy expectations. If we can't find them we'll create them. Our topophilia can never be extinguished or sated.

We are headed for uncharted territory, to places found on few maps and sometimes on none. They are both extraordinary and real. This is a book of floating islands, dead cities, and hidden kingdoms. We begin with raw territory, exploring lost places that have been chanced upon or uncovered, before heading in the direction of places that have been more consciously fashioned. It's not a smooth trajectory, for nearly all of the places we will encounter are paradoxical and hard to define, but it does allow us to encounter a world of startling profusion. As we will quickly discover, this is not the same thing as offering up a rose-tinted planet of happy lands. Authentic topophilia can never be satisfied with a diet of sunny villages. The most fascinating places are often also the most disturbing, entrapping, and appalling. They are also often temporary. In ten years' time most of the places we will be exploring will look very different; many will not be there at all. But just as biophilia doesn't lessen because we know that nature is often horrible and that all life is transitory, genuine topophilia knows that our bond with place isn't about finding the geographical equivalent of kittens and puppies. This is a fierce love. It is a dark enchantment. It goes deep and demands our attention.

The forty-seven places that make up this book are here because they each, in a different way, forced me to rethink what I knew about place. They have not been chosen for being merely outlandish or spectacular but for possessing the power to provoke and disorient. Although they range from the most exotic and grandest projects to modest corners of my own hometown, they are all equally capable of stimulating and reshaping our geographical imagination. Together they conspire to make the world seem a stranger place where discovery and adventure are still possible, both nearby and far away.

Note: Where possible, I have added Google Earth coordinates for the approximate center or location of each place. These coordinates are consistent with each other but cannot be claimed to be exact, in part because they may change each time Google Earth is updated. No coordinates have been given for historical places or places that are mobile.

LOST SPACES

Sandy Island

19° 12' 44" S, 159° 56' 21" E

The discovery of the nonexistence of places is an intriguing byway in the history of exploration. The most recent example came in 2012 when an Australian survey vessel visited Sandy Island, seven hundred miles east of Queensland, and found that it was not there. This despite the fact that a stretched-out oval fifteen miles long and about three miles wide had been on the map for almost as long as people had been charting these seas.

Breakers and sandy islets were first sighted here by a whaling ship called the *Velocity* in 1876. A few years later Sandy Island got a mention in an Australian naval directory. In 1908 its inclusion on a British Admiralty map of the area lent it even more legitimacy. But its dotted outline on this chart shows that it was being identified as a potential hazard and hinted that it needed further exploration. Four years earlier, in 1904, the *New York Times* had covered the story of an American cruiser, the USS *Tacoma*, that had been sent to verify "Hundreds of Illusions Charted as Land" within the "American Group," a chain of islands supposedly located midway between the

United States and Hawaii. Their existence was given weight by the claim of Captain John DeGreaves, "science advisor" to King Kamehameha of Hawaii, that he had picnicked on one of them in the company of the famous "Spanish dancer" and mistress of King Ludwig I, Lola Montez.

Unfortunately both the islands and the picnic turned out to have been wishful thinking on the part of the captain. The *New York Times* story explained why, aside from tall tales, the oceans were still littered with cartographic blunders. "Long dark patches or bright, yellowish patches, which at a distance give the mariner the impression of shoals," "a rip tide mistaken for breakers," and even the back of a floating whale have been enough to start a new myth. In lonely parts of the sea, where information is at a premium and corroboration or refutation rare, the thinnest evidence "will live for a time on a chart with the embarrassing letters 'E.D.' opposite the entry, meaning its existence is doubted."

Because land is looked and hoped for by sailors, the faintest signs have been seized upon. Far from being doubted, Sandy Island's credentials became ever more watertight. Having been inked in on an authoritative chart, it acquired the status of a known fact and its myth was transmitted deep into the twentieth century and beyond. It was included in maps produced by the National Geographic Society and *The Times* of London, and no one complained or even noticed. It was also, apparently, captured by the satellites that many imagine are the sole feeders for Google Earth. Dr. Maria Seton, who led the Australian survey team, explained to journalists that although the island is on Google Earth as well as other maps, navigation charts show the water to be 1,400 meters deep in the same spot: "so we went to check and there was no island. We're really puzzled. It's quite bizarre."

On November 26, 2012, Google Earth blacked out Sandy Island and later stitched over the spot with generic sea. Today on Google Earth the place where Sandy Island once was is crowded with dozens of photos uploaded by map browsers. Unable to resist the creative possibilities, they have scattered the ex-island with images of clashing dinosaurs, moody urban back streets, and fantastical temples.

The story of the disappearance of Sandy Island was a minor global sensation. If Sandy Island doesn't exist, then how can we be certain about other places? The sudden deletion of Sandy Island forces us to realize that our view of the world still occasionally relies on unverified reports from far away. The modern map purports to give us all easy access to an exhaustive and panoptic God's-eye view of the world. But it turns out that ventures such as Google Earth are not just using satellite photographs. They rely on a composite of sources, some of which are out-of-date maps.

Even before 2012 some people already knew that Sandy Island didn't quite live up to its name. It lies in the terrestrial waters that extend hundreds of miles around the French "special collectivity" of New Caledonia. But some decades ago Île de Sable, the French name for Sandy Island, was quietly dropped from French maps, and it doesn't feature on a French hydrographical office chart drawn up in 1982. A 1967 map of the area produced in the USSR also doesn't include it. What is clear is that not everyone is using the same sources. However, this does not imply that the French or Soviets have been consistently better informed than everyone else. The 2010 Michelin map of the world includes Île de Sable, and the news of its nonexistence was as much a surprise to the French public as it was to the rest of the world. Following the Australian nondiscovery of Sandy Island, *Le Figaro* announced, on December 3, 2012, that "Le mystère de l'île fantôme est résolu."

But this isn't just a technical story about mismatching geographical data. Why should it matter to anyone that a sandy strip thousands of miles away, somewhere hardly anyone had ever heard of before, turns out not to be there?

It matters because today, although we live with the expectation that the world is fully visible and exhaustively known, we also want and need places that allow our thoughts to roam unimpeded. The hidden and remarkable places are havens for the geographical imagination, redoubts against the increasingly if not exhaustively all-seeing chart that has been built up over the past two hundred years. The 1908 inclusion of Sandy Island on an Admiralty chart was a clumsy error, a mistake not typical of the period. Far from dotting the globe with fabulous islands, the naval powers of the nineteenth and early twentieth centuries remorselessly tracked down any and all such rumors and either confirmed or disproved them. As a result, the 1875 revised Admiralty Pacific chart discarded 123 unreal islands. The *New York Times* story from 1904 led to the confirmed nonexistence of a cluster of islands south of Tasmania called the Royal Company Islands. After vessels were sent to investigate, these islands, like so many before them, were removed from the map. The American ships were doing their bit for modernity: eliminating doubt, attaining panoptic knowledge. Yet modernity also gives us the self-questioning and self-doubting consciousness that permits us to understand that we lose something in its attainment. As the clutter of outrageous, fantastical photographs that today occupy Sandy Island's place on Google Earth suggests, Sandy Island's disappearance established it as a rebel base for the imagination, an innocent and an upstart that managed to escape the vast technologies of omni-knowledge.

The story of Sandy Island might suggest the need for a survey of undiscovered islands, places once thought to be real but later found not to be. However, it turns out it's already a crowded market. From early titles like William Babcock's *Legendary Islands of the Atlantic* (1922) to comparatively recent studies such as oceanographer Henry Stommel's *Lost Islands: The Story of Islands That Have Vanished from Nautical Charts* and Patrick Nunn's *Vanished Islands and Hidden Continents of the Pacific,* we have an extensive catalogue of the world's nonexistent islands. Some of these studies focus on mariners' mistakes, of which there seem to have been plenty. Others, like Patrick Nunn's, combine the legendary with the scientific. Nunn's interest is in how indigenous legends of lost islands found in many Pacific island communities fit into and inform the environmental history of the area. It turns out that "legendary islands" can sometimes be explained by reference to changing sea levels and seismic activity. Ancient topographical changes are recorded in local myth and lore. A similar link has been made in other parts of the world, most famously in the legend of Atlantis.

Interest in phantom places like Sandy Island is growing. In part this is because such "nondiscoveries" are now so scarce: it is unlikely that many more islands of any size will be "unfound." But there are still plenty of shifting and potentially doubtful phenomena out there, including cartographic "facts" like the shapes of nations, borders, mountains, and rivers, that will continue to disturb our geographical certainties. The truth is, we want to have a world that is not totally known and that has the capacity to surprise us. As our information sources improve and become ever more complete, the need to create and conjure new places that are defiantly off the map also grows.

Leningrad

St. Petersburg wasn't forgotten when it was renamed Leningrad in 1924. It had already changed its name once, to the more Russian-sounding Petrograd, in 1914. But one of its sons, the poet Joseph Brodsky, thought it would always be Petersburg. In his 1979 "Guide to a Renamed City," he observed that its citizens carried on calling it "Peter" and that "Peter I's spirit is still much more palpable here than the flavor of the new epoch." Twelve years later the city became St. Petersburg once more. But Leningrad too will not go quietly. It may have been taken off the map, but that doesn't mean it's gone.

In *The City and the City,* China Miéville's allegory of antagonistic cities that literally cohabit the same space, the inhabitants stay culturally pure by "unseeing" each other and the other place. But the temptation to look is strong, preying on their minds and dominating their every step. This is also true of places that have been replaced and renamed; they manage to be both ghostly and alluring. It is surprising that we haven't grown blasé about such changes. Over its two millennia the ancient Bulgarian city of Plovdiv has seen twelve such changes. Renaming places became a talisman of

progress in the twentieth century. Everything from villages
to countries was rebranded, a seemingly simple act that often had profound consequences for their inhabitants. Some
involved imposing a new ethno-national identity on an old
place. When the Ottoman Empire became "Turkey" in 1923,
and Siam "Thailand" in 1939, loosely defined, multiethnic categories were turned into ethnically exclusive ones. Overnight,
citizens who were ethnically non-Turkish or non-Thai lost
their homeland; they became anomalous and therefore very
vulnerable.

Thai and Turkish nationalists claimed that Siam and the
Ottoman Empire were ripe for renaming. The Ottoman Empire was defunct, and "Siam" appears to derive from a Hindi
word for the region. Ethnic Turks and Thais saw little reason to value the old labels, partly because they were the win-
ners in what appeared to the outside world as a process of
indigenization. But things are rarely so simple. The replacement of "Smyrna" by "Izmir" in 1930 records the expulsion of
the city's Greek population and its rebirth as an ethnic Turkish city. The total disappearance in 1946 of "East Prussia" into
East Poland and the Soviet exclave of Kaliningrad was also an
act of revenge and ethnic cleansing. For hundreds of years this
eastern outpost of Prussia had been predominantly German.
In a few years the Germans were gone, fleeing west from the
Red Army or expelled by Stalin. Yet what Max Egremont calls
"the whispering past" of Prussia keeps coming back. Plans to
return to the exclave's main city, also called Kaliningrad, its
old German name of Königsberg—a name that reminds people of philosophers, monasteries, and castles rather than Soviet troops—keep being raised then shelved.

Although Petersburg's communist past is widely reviled,

refuses to curl up and die. Something too important is buried there: everyday struggles and extraordinary dramas. Leningrad's history makes Petersburg seem shallow. Petersburg was an imperial new town built on the Baltic coast in the eighteenth century by Peter the Great and given a foreign, Dutch-sounding name, Sankt-Peterburg. It faced Europe, the future, and high culture, and away from Russia and its stolid peasantry. Objecting to the displacement of Leningrad by this older but alien rival, the Leningrad-born novelist Mikhail Kuraev used the pages of a Russian literary magazine to claim, "Three hundred years ago the name Sankt-Peterburg sounded to the Russian ear the way Tampax, Snickers, Bounty, and marketing sound to us today." Kuraev regards Petersburg as an "internal immigrant in its own motherland" but Leningrad as authentically Russian.

Leningrad has earned its place in Russian memory: it is soaked in patriotic and revolutionary blood. It was here that nine hundred days of siege were endured during the Second World War, when a starved people defended and then rebuilt their city from the rubble. It was Leningrad that was awarded the status of "hero-city" by Stalin. Even the city's Nazi attackers were impressed, and not just by the grim determination of its citizens. Despite its revolutionary credentials, Leningrad was also a center of alternative thinking. The Leningrad Affair of the late 1940s and early 1950s saw many local Communist Party leaders executed or banished as Moscow sought to root out anti-Stalinism.

It is fitting that a Russian exile born in Leningrad, Svetlana Boym of Harvard University, should become an expert on nostalgia. In *The Future of Nostalgia* she offers a complex, sympathetic portrait of the many ways "the duel between the two cities Leningrad and Petersburg continues." Boym is especially drawn to Leningrad's bohemian side found in the city's cafés and suggests that Leningrad lives on as a kind of alternative or second city, recalling critical "potentialities that have not yet been realized." She is offering us an upbeat message about all those drowned memories, packaging them as a resource for a more liberal city.

But I suspect that turning the usurped city into a bohemian subtext beneath its rival is just another form of forgetting. There appear to be plenty of people in Petersburg whose regret for the death of Leningrad has little to do with this minority political identity. They miss the clear sense of order, the broad social safety net, the respect for the old, the slower pace, the dignity and valor. It may also be the accretion of attachments that keeps Leningrad alive. After all, when it was abolished in 1991 Leningrad had existed for sixty-seven years.

Leningrad may never have quite become the nonconformist city once dreamed of in its alternative cafés, but it was a place of long ordinary years as well as enormous sacrifice. By comparison, today's Petersburg looks like what the Leningrad-born poet Aleksandr Skidan calls "museumification under the open sky." In snuffing out Leningrad, some kind of justice has been dealt to the victims of Soviet communism. But the same gesture also erases these victims and myriad prosaic memories.

Leningrad clings on. The world's first Lenin statue still stands in Lenin Square, although vandals dynamited a large hole in his backside a few years ago. Another of the city's Lenin statues was almost sliced in two in another bomb blast in 2010. Yet the statues were not demolished but repaired. Wiser citizens know that Petersburg is also Leningrad, that the two must somehow live together. It is not love or even respect that Leningrad needs, but acknowledgment. Like so many other renamed places of the world, the city's former self appears more troublesome, but also more interesting, and sometimes more alive, than its replacement.

Arne

50° 41′ 39″ N, 2° 02′ 29″ W

Arne is an example of a sacrificed place. The village, which sits on a small peninsula sticking out into the English Channel, was evacuated in 1942. Close to where the village stood a decoy factory was built, designed to lure German bombers into dropping their loads short of the Royal Naval Cordite Factory, a sprawling munitions complex a few miles north at Holton Heath.

Decoys were widely employed throughout England during the war. Many were far more elaborate than Arne, since diverting bombers from cities required complex operations. After an air raid on Coventry in November 1940, work started on building massive "Starfish" decoys outside nearly all major urban areas, whose purpose was to fool pilots into thinking they were flying directly over a burning city. By January 1943, more than two hundred Starfish sites had been built. In the early days, tons of random combustible materials were thrown at the task, but as the war progressed the decoys became more sophisticated. Steel tanks, troughs, and pipes were used, and fuel was poured, sprayed, or trickled at timed intervals — an entire symphony of pyrotechnics orchestrated from a control bunker. The brightest was known as the Boiler Fire and periodically released oil from a storage tank into a heated steel tray where it vaporized. Every so often water was dropped into the tray, producing huge flashes of white-hot flame that leaped as high as forty feet. A typical Starfish site might contain fourteen Boiler Fires and burn through twenty-five tons of fuel every four hours.

The Starfish were a great success, and by June 1944 decoy sites had been attacked on 730 occasions. Although Britain's cities were heavily bombed, the fact that they were not obliterated is in large part a testament to the work of the decoys. In drawing the high explosives and incendiaries onto themselves, they were responsible for saving the lives of thousands of people. Today a few of the control bunkers remain but the decoy sites themselves have disappeared, sinking back into the surrounding landscape.

The decoy at Arne consisted of a network of tar barrels and pipes carrying paraffin, which could be set alight to make it appear from the air as if buildings were on fire. The strategy worked. While hundreds of bombs were dropped on Arne, the factory at Holton Heath escaped almost untouched.

Today Arne is a peaceful and beautiful place. After the war the village was permanently abandoned and lay derelict until the late 1950s. In 1966 the Royal Society for the Protection of Birds took over the site and renovated the remaining buildings, including a thirteenth-century church and a former Victorian-era school. The deserted village has become a feature of a nature park that covers much of the peninsula. The bomb craters have become wildlife havens and the gun emplacements are smothered in weeds. Troops of green-clad

bird watchers line up in the parking lot, shouldering tele-
scopic sights, on the hunt for rare local species such as the
tiny Dartford warbler.

The military landscape may have been swallowed up but
it has not been digested: the layering of such a peaceful landscape over such a violent one is unnerving. The sandy, flower-
covered heath that dominates Arne is "preserved" and "protected," but the sense of abandonment lingers on and disturbs
the security and comfort that those labels imply. The memory
of desperate violence and loss has been veiled here by new associations, but that process also serves to make modern-day
Arne appear fragile and temporary.

Arne is one of 250 abandoned villages in Dorset. Some
are little more than medieval bumps on the ground, but others are more recent. The most famous twentieth-century example is Tyneham, a village along the coast from Arne that
was evacuated in 1943 to allow the army to practice live firing.
In the buildup to D-day much of this coast was designated
a battle range because of its similarities with the Normandy
beaches. In Tyneham a notice was pinned to the church door:
"We have given up our homes, where many of us have lived
for generations, to help win the war to keep men free. We
shall return one day and thank you for treating the village
kindly." But the villagers never came back. Today Tyneham
is a collection of derelict stone houses that still sits within a
military firing range and, like much of this part of Dorset, remains under the control of the British Army.

A recent archaeological survey concluded that the Arne
decoy "had been given over to agricultural use and no features
of the decoy survive." I spent several hours combing the fields,
woodland, and reeds at the spot where the decoy is said to

have been sited and didn't find much, at least not much that I understood. There were a few huge bomb craters and two pairs of military gates, broken off their hinges and engulfed with bindweed, but there was little else other than some large wooden pegs, a lot of badly decayed dyed green wood, and a sickly orange tinge to patches of leaf matter. I'm not sure why I expected my amateurish foraging would turn up anything more exciting when archaeologists had already written the place off. I was drawn to the idea that somewhere that burned so brilliantly and dangerously would leave more than just a stain. But all I can claim to have found is an uncertain and quiet trace. My presence disturbed some deer that bolted off into the neighboring wetland. No one had been here for a long time. Arne is a place of lost drama and disconcerting stillness. Its past of fires, bombs, and evacuation gives its present tranquility a disturbing edge. I left the wood and made my way along a narrow causeway between reed beds to a small rocky island covered with gorse bushes. It was a warm, sunny day and I could have lain down and surrendered to the birdsong, but Arne had disoriented me and left me restless.

Old Mecca

Some cities are lost while their inhabitants continue to go about their everyday lives. Over the past two decades around 95 percent of the ancient city of Mecca has been demolished. The city has been rebuilt as a set of broad roads, parking lots, and hotel and retail blocks. Even the English name has been upgraded: the Saudis now prefer to call it Makkah.

Today the skyline is dominated by the massive Makkah Clock Tower hotel, a giant Soviet-style Big Ben that soars above the holiest sites in Islam, the Kaaba and the Grand Mosque. The innocent obtuseness of this profanity is caught, albeit unintentionally, on the hotel's website: "Makkah Clock Fairmont Hotel Tower stands out modestly and respectfully as Makkah's second mark of distinction with an outstanding clock that can be seen from 17 kilometers away."

The Mecca building boom has been driven by the need to provide new facilities for the more than three million pilgrims who come every year, but also by the hard-boiled nature of Saudi iconoclasm. For many centuries Islam has prohibited the creation of images of humans or animals, but the kingdom's puritans have joined forces with property developers to enforce a far more wide-ranging agenda that targets all old buildings and monuments.

As many cities are still learning, sweeping away the past deprives the world of more than just rare and beautiful landscapes. Planners and developers also remove the memories, stories, and connections that hold people together, socially as well as individually. Turning complex, diverse places into shallow, simple ones creates a more culturally vulnerable population, an unrooted mass whose only linking thread lies in the ideology that is fed to them from above.

This process was well understood by those communist regimes that undertook similar mass demolitions in the past. In *The Destruction of Memory,* a profound book on the politics of urban reconstruction, Robert Bevan chronicles the almost fetishistic desire for pulling places down that accompanied the creation and maintenance of state communism. He explains the failure of attempts "to persuade Mao that the new

Beijing should be built adjacent to the ancient, sacred city" by pointing out that, for Mao, molding the people to his will demanded the death of the old city. The "obliteration of the past was as much a consideration as the building of the new."

In the face of puritanical ideologies, whether political or religious, the past takes on a subversive and unruly quality. Old photographs of Mecca show a labyrinthine city, sets of courts, mosques, and alleys piled up on the low hills, a density of periods and influences. Today what little remains of this complex past survives by luck or because it is just too prominent to pull down. Even the most important mosque in Islam, the Grand Mosque, has been under assault. The Saudi antagonism to old buildings is heavily tinged with sectarianism and targets the physical proof that Islam was once practiced in the city in different ways. The long presence of the Abbasid caliphate and then the Ottoman caliphate and empire has been almost entirely expunged, and in the past few years the ancient Abbasid and Ottoman sections on the eastern side of the Grand Mosque have been pulled down.

Continuing its unintended taste for paradox, the website of the Makkah Clock Tower hotel picks out the handful of buildings that survived the kind of destruction caused by the hotel's own erection as must-see destinations for visitors. Thus visitors are encouraged to take in Qasr as-Saqqaf, touted as "one of the oldest buildings in Makkah and an ideal representative for the traditional architectural design." Another must-see is the Qasr Khozam palace, which, having been "built over 80 years ago," apparently counts as early heritage in modern Mecca.

Yet any survival from the past irritates the purists, who want complete control of the city. Apologists for Saudi turbo-iconoclasm claim that it is entirely religiously inspired. The country's highest religious authority, Sheikh Abdul-Aziz Ibn Baz, issued a fatwa in 1994 decreeing, "It is not permitted to glorify buildings and historical sites . . . Such action would lead to polytheism." The sheikh was repeating a mantra that has been stamped on this land for a little over two hundred years. The Wahhabis, the Islamic faction to which the Saudi dynasty belongs, took control of Mecca and Medina in 1803. From the first they were intent on tearing down visible associations with other, older, and less puritanical varieties of Islam. The mausoleums and mosques cared for and often elaborately embellished by the Ottomans were a particular target, including the tomb of Muhammad himself. The tussle between the Ottomans and the Wahhabis over whether to revere or destroy the physical heritage of Islam entered another phase when the Ottomans managed to take back Mecca and Medina between 1848 and 1860. But by the century's end the holiest places in Islam were firmly back in the hands of a sect that regards respect for Mecca's past as idolatrous.

The destruction of old Mecca goes hand in hand with the ban on non-Muslims entering the city, as well as the center of Medina. Both are attempts to cleanse the city of historical complexity. The road signs on the freeway into Mecca spell it out: "Muslims Only." A side road taking cars away from town is marked, "Obligatory for Non-Muslims." Of course there are plenty of bans on nonadherents entering religious sites around the world, including non-Mormons and non-Hindus entering Mormon and Hindu temples, at least during services and rituals. But the scale of the ban in Mecca, which prohibits five-sixths of the world's population from entering not just one building but an entire city, makes it unique. The Qur'an has a verse that instructs, "Truly the idolaters are unclean; so let them not, after this year, approach the Sacred Mosque." It's not a prohibition on entering Mecca but on entering the Grand Mosque. The citywide ban is another invention of the Wahhabi Saudis. Ironically, before they seized the city, the Wahhabis themselves were deemed heterodox and banned from its holy places by the city's sharif, or holy steward.

Yet if the motivations are entirely religious, why have both secular and religious historical buildings been targeted for demolition? It's worth recalling that communist regimes also claimed to have purely ideological motives for knocking down old buildings, but their actions can today be seen as more to do with securing power and profits. Mecca is a boomtown with a guaranteed and growing flow of cash-rich pilgrims. "All the top brands are flocking here," says John Sfakianakis, the former chief economist of Banque Saudi Fransi. Starbucks, The Body Shop, Topshop, Tiffany & Co., Claire's accessories, and Cartier are just a few of the labels that are benefiting. The iconoclasm inflicted on Mecca is providing the perfect environment for the growth of consumerism. Nothing stands in the way of spending: no signs or symbols of the kind of slower, less frenzied, and more heterogeneous way of life that must have existed here before history was erased.

The destruction of old Mecca and the ban on non-Muslims impose a singular vision of the city's past and future. They also provoke nostalgia for that lost diversity. Like Leningrad, as old Mecca recedes from reality it emerges as a place of fantasy and critique. In an ancient city like Mecca the glaring absence of the past becomes its own form of presence, something intangible but also something permanent and important that lies inside the city's story and which can never be entirely extinguished.

New Moore

21° 37′ 00″ N, 89° 08′ 30″ E

Dramatic storms and floods can reshape a landscape overnight. The island of New Moore emerged a few kilometers out to sea in the Bay of Bengal after Cyclone Bhola in 1971, and was built up from the soil, sand, and stones that had been tumbled down the many rivers that make up the vast braided landscape of the Ganges Delta. The island grew quickly, eventually reaching 3.5 kilometers in length and 3 kilometers in width. A handful of Bangladeshi fishermen lived on the island during the dry season but it was otherwise uninhabited. Apart from a cluster of mangrove trees, there was little to keep New Moore in one place. Its size and shape shifted with the seasons and the tides.

New Moore could hardly have chosen a more delicate spot in which to take residence, since the Hariabhanga River from which it sprang forms the border between India and Bangladesh. As soon as it crested the water, both countries claimed it as their own, each giving it a different name. For the Indians it was New Moore Island, which remains its most commonly used name, but for the Bangladeshis it is South Talpatti. Either way it was a rich prize. India and Bangladesh have overlapping claims on the Bay of Bengal and its oil and gas reserves. Being able to claim a new island so far out to sea would allow one of them to extend its territorial waters over lucrative seabed.

As a signal of intent, the Indian Border Security Force installed a billboard on the island in 1978, complete with a map of India and a picture of the Indian flag. The Indians raised the temperature again in May 1981 when they temporarily stationed troops on the island and ran a real flag up a real flagpole. For a while the question of which country this sandy speck belonged to looked as if it might lead to a serious conflict. Both sides, however, hoped that the conclusive shots would come from independent boundary experts. Those experts were tasked with determining how the waters of the Hariabhanga flowed around the island. This arcane information could have been decisive if it could have pinpointed the river's *thalweg,* a German word used in boundary disputes involving rivers that refers to the line of a river's lowest, and hence middle, flow. If the *thalweg* was to the east of the island, then it was India's; if it was west of the island, it belonged to Bangladesh.

As it was, the *thalweg* proved hard to determine, and delays set in. But before the problem was sorted out, New Moore began to disappear, and in March 2010 it was fully submerged. The last photograph of the young island shows the topmost branches of drowned trees clawing the waters.

Rising sea levels are creating new shorelines at a rate that is outstripping governments' abilities to respond. Since 2000 the waters of the Bay of Bengal, which were already rising, have been getting higher, quicker. The bay now sees a rise of about five millimeters a year. In a low-lying region, subject to sudden inundations, that's a significant increase. For some,

New Moore looked like a political problem caused by nature and solved by climate change. The *Christian Science Monitor* ran the story as "Global Warming as Peacemaker." In fact, separating out what is natural about either the rise or the fall of New Moore isn't that easy. Only one of the contributing factors falls simply into the category of "natural": the sinking of a tectonic plate. This subsidence is gradually lowering the land under and around the Bay of Bengal and increasing sea levels in the process.

And climate change is making the situation worse, not better. It can be blamed not only for accelerating the cycle of creation and destruction but also for the severity of recent floods. The increased rainfall that created the swollen rivers that, in turn, gave birth to New Moore was the direct result of the warmer seas caused by climate change. Road-building upstream also contributed to New Moore's creation, by triggering landslides that added huge loads of sediment to the river. Unfortunately, deforestation across the region, especially the axing of mangrove trees along the coast, meant that the new sediment did not sink at the shoreline and help defend the coastline but instead was carried far out to sea.

New Moore is not the only island that has come and gone in the Bay of Bengal. On the Indian side at least four other islands have emerged and then vanished. One of these, Lohachara, had a population of six thousand before it went under in 2006. Recently it has been spotted rising again. It seems that neither the appearance nor disappearance of these islands is a one-off event. They are rising and falling with greater frequency, a phenomenon that has been noted at other estuaries across the world. One of the most famous new islands emerged in France in January 2009, after Cyclone Klaus hit southwestern France. The Gironde estuary deposited what was soon named L'île Mystérieuse seven miles out to sea. Covering 250 acres at low tide, L'île Mystérieuse was caused by many of the same processes as New Moore. The waters of the Bay of Biscay are not rising as fast as those of the Bay of Bengal, and with luck L'île Mystérieuse may be around for longer than New Moore, since new islands off lowland coasts can be very useful. Their environmental worth doesn't lie in pushing back territorial claims but in protecting coastal areas from storms and inundation. They could also provide additional land for overpopulated nations. In a world where coastline change is speeding up and becoming more unpredictable, such outcrops should be given a helping hand. New Moore could be raised up and bulked out and held in place with mangrove trees. It would be an inexpensive task, at least when compared with the construction of entirely new islands.

As has become clear, sea level rise doesn't augur a dawn of fun holiday islets but a wearisome struggle to protect low-lying land. Projections suggest that much of the Bay of Bengal, from Kolkata in the west to Myanmar in the east, will soon be under water. Initial predictions by the Intergovernmental Panel on Climate Change warn that Bangladesh will lose about 17 percent of its landmass by 2050. More recent work by the Dhaka-based Center for Environmental and Geographic Information Services indicates that much of this land will not be lost permanently but seasonally, during monsoon. Either way, it's a growing problem in a country where every hour an average of eleven people already lose their homes to rising water.

New Moore represents far more than a dispute over territorial waters. It is a serendipitous and heavy nudge toward a more proactive approach to sustaining coastal islands. Living in a delta-based country, Bangladeshis are used to the idea that islands can come and go, and they have the skills to hang on to them. Another recent arrival shows how it is done. Nijhum Dwip, or Silent Island, emerged in the early 1950s. Although much of the island is regularly flooded, it has now been stabilized and consolidated, largely thanks to the planting of mangrove trees. More than ten thousand people live there, along with deer, monkeys, and a planned tiger sanctuary. In 2001 it was designated a national park.

Given assistance, new islands like New Moore can become viable places. It's true that we don't know if land-building activities can outpace the rise of the sea, and pessimistic forecasts suggest that the only long-term solution for many coastal areas is abandonment. But for the time being there is no reason to run for the hills. Rather, we need to extend our idea of what local conservation consists of. Twenty-first-century conservation will need to include not just protecting species and ecosystems but island-making too. Islands do not need to be left to sink; they can be managed and sustained. With help, New Moore could rise again.

Time Landscape

40° 43' 37" N, 73° 59' 58" W

At the corner of LaGuardia Place and West Houston Street in New York is a rectangle of land, fenced in and inaccessible to the public, that since 1978 has been given over to lost nature.

This quarter-acre plot was planted by the artist Alan Sonfist with species native to the area. Red cedar, black cherry, and witch hazel, along with groundcover of Virginia creeper, poke-weed, and milkweed—the kind of flora that would have been found in the city before the seventeenth century.

Time Landscape was the first major work to emerge from ideas that Sonfist had been nurturing for some while. In a manifesto published in 1968, "Natural Phenomena as Public Monuments," he called for environmental equivalents of war memorials. Such places would become monuments to vanished landscapes, places of reflection that record and remind us of "the life and death of natural phenomena such as rivers, springs, and natural outcroppings."

Time Landscape is designed to "be a reminder that the city was once a forest." It is also a more personal reminder. Sonfist admitted in a recent interview with John Grande that much of his work "began in my childhood when I witnessed the destruction of the forest, walking in the Bronx." Yet in its completed state *Time Landscape* poses some difficult questions about the defense of nature's lost places. For *Time Landscape* is constantly being invaded by alien, post-Colonial weeds like morning glory and sow thistle. Sonfist said he is not bothered, arguing that "this is an open lab, not an enclosed landscape" and that he always meant there to be an interplay between species.

Yet if that is the case, then *Time Landscape* is a rather hollow memorial. It is precisely its exacting evocation of the past that makes it different from any other bit of green space in the city. It's no surprise that New York City's Department of Parks and Recreation, which now manages the site, is less blasé about alien weeds. All such invasive species are cleared out at intervals. *Time Landscape* has become a unique if rather low-key asset in the department's program called Greenstreets, which it says is designed to convert "paved street properties, like triangles and malls, into green lawn." The department wants to preserve *Time Landscape* as art. Thus it has become subject to another type of preservation, another attempt to stay the corruption of time.

Removing the weeds from *Time Landscape* maintains it as past art. Without all that grubbing up, its temporal direction would get much harder to read: it would be less clear whether it was pointing backward or forward. Critics say *Time Landscape* has been "museumified," that it's now a dead place and of little public benefit. In fact, its layers of preservation have combined to make it ever more complex and disconcerting. *Time Landscape* has gotten weirder, for it now confronts us with an uncomfortable paradox: as we try to revere nature, it slips through our fingers, leaving us holding something we never expected, something unnatural.

The city is a place where nature is excised and then mourned, killed off then raised from the dead, only to be entombed in caged-off spaces of floral tribute. The weeds that infest *Time Landscape*'s sepulchral landscape are pulled up and stuffed into black plastic garbage bags and removed for incineration. They form their own kind of monument, off to the fires, our revenge on the revenge of nature, enacted again and again. The carefully maintained remnants of nature that remain are too anemic to evoke a fertile or meaningful past, even as they secure *Time Landscape*'s status as a memorial to both past nature and past art.

Time Landscape's protocol of purity is echoed across

countless parks and gardens but also in the kind of environmental or land art that tends to get commissioned in cities.
A lot of land art creates disorientingly human places within
large natural landscapes: a straight stone path amid a chaos of
boulders, a spiral jetty thrust into a remote lake. But for artists working in cities the temptation to confront paved streets
with pure nature seems irresistible. Apart from *Time Landscape,* the best-known work in this genre seen in New York
was *Wheatfield—A Confrontation,* a two-acre vacant lot in
downtown Manhattan that was planted with wheat by Ag-
nes Denes in 1982. It was a more political piece than *Time
Landscape.* The fertile field and the one thousand pounds of
wheat yielded were symbolic of the hunger caused by Wall
Street's "misplaced priorities." But the golden grain and the
simple moral message were also designed to contrast with the
corrupt and fallen city. This was another place made pure by
nature.

Denes's *Wheatfield* was soon reaped, and wasn't around
long enough to get drawn into difficult debates about land
use or to start looking outdated. *Time Landscape* suffers a
different fate. The *Village Voice* reported the atmosphere of
one weeding and cleanup day turning slightly sour as the director of a local community alliance declared that "the time
has come for something new" and that "Time Landscape is a
piece of '80s art," all "within earshot of the artist." It's true that
over the past couple of decades the pursuit of prelapsarian
eco-art has gone out of fashion and a fascination with weed-
infested urban decay has taken root (see "The Archaeological
Park of Sicilian Incompletion," page 119). An influential essay
by John Patrick Leary on the "exuberant connoisseurship of

dereliction" labeled the trend "Detroitism," because for artists and photographers, that city has become "the Mecca of urban ruins."

Time Landscape and Detroitism are very different starting points, but they both converge on a central worry for urban civilization: How can we live without nature? What do we become without it, or what can we pretend to be? Sonfist never claimed to have the answers, and what *Time Landscape* means has long since escaped his control. Today it is a troubled and paradoxical place but also somewhere that hints at the remorse for lost nature that lies just beneath the surface of even the shiniest and blankest cityscape.

The Aralqum Desert

44° 45' 37" N, 62° 09' 27" E

The Aralqum Desert is too new, too large, and its outline too changeable to be on any maps. It's a desert that used to be called the Aral Sea. The new name is gaining favor, although it's not quite as exotic as it sounds. *Qum* is Uzbek for "sand."

The map captioned "Geography: Physical" is usually seen as an impassive affair when compared to "Geography: Political." We are used to the latter requiring regular updates but continue to imagine that the physical outlines and natural features of the planet are slow-moving or even rock-solid. The love of "natural places" is, in part, built around the conviction that, unlike our fragile settlements and fickle borders, they are self-reliant and age-old. It's an outdated perspective, as New Moore (see page 20) demonstrates, and encourages a belief that natural systems can always cope with change; that when one set of flora and fauna die out, a new set will happily move in. The Aralqum is a natural place, an empty desert, but also an unnatural one that shows that organic adaptation can no longer keep pace with human impact.

It's also a place of disconcerting memories. The Aral Sea was once enormous. At 426 kilometers long and 284 kilometers wide, it was the fourth-largest lake in the world. Any schoolchild tracing her finger across the map of Central Asia will still find it and pause and wonder how such a big blue shape could have formed so many miles from the ocean. It was once called the Blue Sea and was first mapped in 1850. Soon the Aral Sea was supporting several fishing fleets and a cluster of new villages, and by the middle of the last century it was fringed by nineteen villages and two large towns, Aralsk in the north and Muynak in the south. Today these towns' harbors are many miles from water.

The Aral Sea was fed by one of the longest rivers in Central Asia, the Amu Darya, which flowed north for 1,500 miles to spawn an island-flecked delta. Along with the Syr Darya, which fed the Aral's northern shore, the Amu Darya pumped the Aral Sea full of fresh mountain water. Soviet planners were not slow to see the potential of these rivers to feed cotton and wheat irrigation systems. Starting in the 1930s, huge channels were constructed, diverting water from both the Amu Darya and Syr Darya and spreading it out over millions of acres of fertile land. One of the Soviet Union's most eminent experts in desertification, Professor Agajan Babaev, explained in 1987, in an article for a Soviet economics magazine, that "the drying up of the Aral is far more advantageous than preserving it." Even more oddly, he also concluded that "many scientists are convinced, and I among them, that the disappearance of the sea will not affect the region's landscapes." The death of the Aral Sea was not only foreseen but actively pursued.

As the Aral Sea began to shrink, in the 1960s, the irrigation continued, the volume of water drained off the rivers only peaking in 1980. Without the rivers' infusion of fresh water, many of the Aral's shallowing pools became almost as salty as the ocean. A new dusty and denuded landscape emerged. Windblown pollutants turned the area into one of the world's unhealthiest places to live, and infant mortality rates shot up along with respiratory diseases. The loss of the Aral Sea also had an impact on the climate. Such a large body of water had long kept the land warmer in winter and cooler in summer. With its disappearance came more extreme and more destructive localized weather systems.

Since 1960 the Aral Sea has shrunk by more than 80 percent and its water volume has fallen by 90 percent. The size and shape of the Aral Sea on recent maps varies enormously: sometimes it is represented quite accurately, as fragmented and shrunken, but it is still common to see it portrayed as undiminished and unbroken. With cotton production still an economic priority in Kazakhstan and Uzbekistan, and no real prospect of rehabilitation in the foreseeable future, it is time the Aral Sea was removed from the world's maps and the "Aralqum Desert" inserted.

Visitors to the Aral today are faced with whipping winds across a barren plain. It is littered with bleached seashells and the remnants of scavenged boats—a desiccated land that stretches to the horizon. The Aralqum Desert is fringed with ghost towns, abandoned fish factories, and rusting boatyards. Barsa-Kelmes, which translates as "the land of no return" in Kazakh, was once the Aral Sea's largest island and used to be a nature reserve, renowned for its eagles, deer, and wolves. Today it is just another dead stump of land. By 1993 it was empty except for one resident, who refused to leave, and a few stubborn wild asses. It seems that the holdout, an ex-ranger named Valentin Skurotskii, was rooted to the island by the fact that his mother was buried there. His body was discovered in 1998, sitting in a chair with his head in his hands.

In Kazakhstan and Uzbekistan people have grown tired of sad tales and bad news about the Aral. Much of the regional news coverage about the Aral over the past two decades has been about the damming and "rebirth" of the so-called Small Aral Sea in the north. The implication is that the rest of the Aral should be abandoned to the sand. The newly built dam that keeps the waters of the Syr Darya in the Small Aral further restricts their flow farther south. In 2008 the president of Kazakhstan, Nursultan Nazarbayev, stood on a new dam near Aralsk and declared that one day the waters would return to the town's harbor. Thanks to the new dams and locks, he may be right. The waters of the Small Aral have risen and become fresher. But it is a meager triumph compared to the loss of the "Large Aral Sea."

The Aralqum is not simply a vast new desert; it is also a huge experiment, the world's largest example of anthropogenic primary succession. Primary succession refers to the development of plant life on land that is devoid of any vegetation. The classic examples are volcanic islands like Surtsey, which emerged from the Atlantic, twenty miles south of Iceland, in 1963. The first plant on Surtsey was spotted two years later, and today much of the island is covered with mosses, lichens, grasses, and even some bushes. While it's a natural process, it's the anthropogenic, or humanly caused, part that turns it into something less predictable. These days most examples of primary succession are caused by humans, and they have nothing to do with volcanism or glaciers. They occur in the wake of the dead landscapes caused by nuclear testing or are found on top of slag heaps or at battle sites or in the cracked tarmac and paving stones of our cities.

These plants seem such doughty invaders that it is easy to assume that, given time, the green world will always grow back and take over. It's early days yet, but at the moment it seems that the Aralqum is suggesting otherwise. The salty, dust-blown, and often poisonous seabed makes conditions very hard for new life. A German team from the University of Bielefeld has studied the limited plant life that is taking root. Along with other experts they predicted that the desert will only be greened by people going in and planting species that are not just salt resistant but can withstand the extreme temperatures and winds of the dry sea floor. Yet 70 percent of the Aralqum is salt desert. To turn it into something living would be an expensive, long-term, and probably thankless task. The Aralqum appears to be showing us that, at least in the short term, nature cannot cope. A problem created by us can only be solved by us, but so far it appears to be beyond us. We have gotten used to seeing natural places as places that can be protected and nurtured, but the story of the Aral Sea indicates a daunting challenge, of moving beyond designating zones of conservation toward rebuilding entire ecosystems and landscapes on a vast scale.

In the meantime, the new desert is sharing its secrets. It

seems this is not the first time that the area has been dry. On the old sea floor Kazakh hunters have found the remnants of a medieval mausoleum along with human bones, pottery, and millstones. Satellite images have also revealed the courses of medieval rivers meandering through the desert. These findings confirm a local legend that the Aral Sea was once land. The area's folklore has since been updated. Now old-timers look forward to a second inundation, a new flood to give them back their blue sea.

Lost places have an uncanny presence in our lives. In a century that has seen the obliteration of so many places, it might be thought that these ghosts would have been exorcised. But that's not how humans work; place means too much to us for its disappearance to ever feel easy or complete.

HIDDEN
GEOGRAPHIES

The Labyrinth

44° 56' 14" N, 93° 12' 03" W

In a world where it is easy to assume that everywhere is fully known and fully charted, places that don't appear on maps become intriguing and provoking. Hidden geographies are the inverse of lost places; they hint at the possibility that the age of discovery is not quite over. The surprising resilience of closed cities and unnoticed uses of existing landscapes challenge us to see ordinary streets in new ways. The underground city provides more intimate hidden places that manage to be both near and far.

Urban exploration took off in the early 2000s. I first knew it was going mainstream when my sixteen-year-old nephew told me he had spent the night in an abandoned mental hospital. He showed me the photos: empty wards full of fallen plaster and upended radiators, grinning teens posing in front of the goofy monsters they had painted on the walls. I didn't ask why he did it, because I already knew. A decade earlier I'd helped set up a magazine dedicated to the experimental geographical wanderings and disorientations known as psychogeography. We called it *Transgressions: A Journal of Urban Exploration*. It ran for only four issues and was full of pur-

posefully perplexing accounts from the geographical avant-
garde. What brought the group together was an understanding of urban exploration as a kind of geographical version of
surrealist automatic writing. Our real-world adventures were
little more than pegs on which to hang our interpretative essays, which usually came with pendulous bibliographies featuring situationists and Magical Marxists. For me it was only
when people like my nephew started going out and laying
claim to hidden parts of the city that I began to understand
that open-air haphazard ramblings can seem very tame when
compared to more purposeful adventures: geographical missions targeted and designed to gain access to forbidden and
unseen spaces in and under the workaday world.

Today this kind of urban exploration isn't, for the most part,
done for the sake of art or politics but for the love of discovery.
The web is filled with message boards for modern urban exploration, where you can find reports from groups in dozens of
cities. New legends are being established by thousands of metropolitan Columbuses. Some of the best-known play spots are
the catacombs and quarries of underground Paris, the dead
subway stops of London, and the abandoned factories and embassies of New York and Berlin, but the nomadic spirit of urban explorers keeps finding new possibilities and taking ever
bolder risks to journey into the map's blank spaces. The burgeoning nature of this scene is reflected in the fact that it has
begun to suffer from internal splits and territorial disputes. At
least some of the discoverers of the hidden city like to think
that they have sole rights to their finds or, at least, that access
must be restricted to an elite clan of fellow travelers.

This story is about the secret world under Minneapolis–

St. Paul, which has been labeled the Labyrinth by urban adventurers. The excitement of exploring the multifarious tunnels and cave systems that make up the Labyrinth was captured by the Action Squad, a band of Twin Cities explorers who specialize in subterranean voyages. After trying plenty of manholes, they found the entrance to the rumored system, a portal that eventually revealed to them seven interconnecting tunnel routes and myriad man-made caves and underground chambers of demolished buildings. Like any other group of pioneers, the Action Squad relished the idea that they were the first to find this lost world, noting on their website the "almost perfect absence of graffiti, explained by the lack of access points achievable by anyone but truly dedicated explorers."

The Labyrinth is hard work, but it offers that mixture of adrenaline rush and breakthrough that makes the effort addictive. "We've spent hours digging tunnels through solid sandstone using butter knives and other primitive tools to bypass barriers that stood in the way of our exploration," recalls a team member on the Action Squad homepage. "We've exclaimed dozens of variations on the theme of 'holy fucking shit!' as we found still more amazing places to explore after thinking we'd already seen it all. God, we *love* that place."

The dedication of the Action Squad and the quality of their finds have drawn other adventurers to the Labyrinth. In an example of urban exploration tourism and homage, a Calgary-based explorer called K.A.O.S. visited the Twin Cities in 2007. "I had to do the Labyrinth. I knew the stories too well," K.A.O.S. writes on a Canadian urban exploration website, adding, "To me it was like stepping into the original UE mythos." After taking a guided tour through some of the

Labyrinth's highlights, the natural caves and under-river passages, K.A.O.S. is left with reverence for the "people who did this for the first time, not knowing whether these tunnels led anywhere, facing the possibility of getting stuck or causing a cave in. That's gotta take balls."

Many of the once hidden places discovered in the first wave of urban exploration have become well known among the cognoscenti, and knowledge about where and how to progress through them grows increasingly commonplace. The shift from an activity shared by a hardy few to a leisure pursuit enjoyed by thousands is a cause of regret among those who want to keep places like the Labyrinth pristine. Another of the early voyagers into this system, university geologist Greg Brick, noted in his local handbook *Subterranean Twin Cities* how, from a few "committed souls," the scene had boomed: "The result was predictable: the subterranean venues, hitherto silent and inviolate, were overrun." Brick has attacked Internet-savvy "point-and-click kids" for despoiling the cities' hidden kingdoms and, in a move that provoked outrage within the Twin Cities underground community, placed a lock on the entrance to one prime site, the Heinrich Brewery Caves.

"I thought that was kind of—pardon my French—but kind of a dick move," complains Action Squad member Jeremy Krans. "It's not his place. None of us go locking things up trying to keep other people out." Krans was talking in 2013 to a reporter from a Twin Cities newspaper that headlined the article "Cave Wars." The story had added bite because the Action Squad claims Brick plagiarized their missions for *Subterranean Twin Cities*. Brick denies this, and a bitter legal dispute has resulted, which has opened unexpected challenges for this formerly carefree community of trespassers. It seems that once their activities become widely known they change in character. Following the routes and finding the places of others, even if those routes and places are illegal and dangerous, may still be an act of adventure, but it is less clear whether it's an act of exploration.

Such thorny issues are likely to become more visible as urban exploration grows. Yet too much emphasis on originality and being the first misses the point for most of its participants, which is about the thrill of discovering the extraordinary in the ordinary city. Bradley Garrett, a geography lecturer at the University of Oxford, who has brought the topic to the pages of geography's scholarly journals, explains that the core values of urban exploration derive from "desires for emotional freedom, the need for unmediated expression," and "associations with childhood play." In 2012 Garrett put his words into practice by climbing up the outside of the Shard, a new London skyscraper, a month before it was finished.

Another explanation comes from Brandon Schmittling, the founder of Survive DC, a kind of citywide game of adult tag based in Washington. "I think people like to believe there's more out there that hasn't been seen," he told *Newsweek*, adding that urban exploration challenges people "to shed their fears about the city."

What is also striking about the urban explorers is their affection for the previously unloved places they discover. They often picture themselves as ragged desperadoes, but their relationship to their sites is actually one of care. They research and document the places they discover with an attention to detail and an offhand but deeply felt respect. Not so much punk Columbuses, perhaps, as urban Alexander von Humboldts, they collect and collate fragments of information in order to create a sense of possibility and celebrate the fact that the mundane world contains within it, or under it, far more pathways and far more fun than we previously thought.

Zheleznogorsk

56° 15′ 00″ N, 93° 32′ 00″ E

In April 2010 two white-coated scientists laid flowers on top of the control rods of a nuclear reactor in Zheleznogorsk, a town founded in 1950 for the sole purpose of making nuclear weapons. For forty-seven years the reactor had been producing weapons-grade plutonium in a city that officially did not exist and was closed to the outside world. The ceremony on the reactor marked the end of an era, and it might have looked like the end of Zheleznogorsk itself, for its ninety thousand residents were nearly all in some way dependent on this one site.

Zheleznogorsk is a grid city of wide boulevards, a place of calm solemnity and perseverance. It was once a secret city. It did not appear on Soviet maps and is still missing from many. For most of its existence it didn't even have its own name and was referred to by a post office box number, Krasnoyarsk-26 — Krasnoyarsk being the nearest big city, forty miles away. It was only in 1992 that its existence was officially confirmed, when President Boris Yeltsin decreed that closed cities could finally be revealed.

Yet Zheleznogorsk is still closed and entry is highly restricted. The hosts of any visitor must submit their request to the security services and the Ministry of Atomic Energy, and even local residents need to get permission to come and go. Surprisingly, Zheleznogorsk remains closed because its residents like it that way. In 1996 they voted to remain shut away from the world. It is at this point that the story of Zheleznogorsk begins to contradict our preconceptions about life in secret places within authoritarian regimes. Closed places and secret cities fitted snugly into the paranoid mindset of Soviet communism, but in a postcommunist era there are other reasons why communities might decide to be cut off from the rest of us. It's not only about hanging on to secrets; it's about holding on to a lifestyle.

Closed cities were once among the best-funded and most prestigious settlements in the USSR, with well-paid jobs that attracted high-achieving technicians and scientists. They were aspirational destinations. The tranquil, kempt character of Zheleznogorsk, with its large park, lakeside setting, and forests and hills, is something its residents want to preserve. They have witnessed what "opening up" has done to the rest of Russia, and they aren't keen to go the same way. Soviet nostalgia hangs heavy in Zheleznogorsk: it's the kind of place that the USSR always promised its citizens. The adulatory website "Zheleznogorsk: Last Paradise on Earth" appears not to be ironic. It's where one local writer, Roman Solntsev, describes the town's appeal as a "wonderful feeling of relaxation, calm and peace of mind." Solntsev goes on to point out the "sharp contrast with the soot-covered, noisy industrial centers and big cities."

Zheleznogorsk is part of a club of approximately forty "closed administrative-territorial formations," which are home to 1.3 million Russians who embrace what might look to the

outside world as something imposed. One ex-resident of another closed city labeled with a box number, Kuznetsk-12,
posting on a chatroom about why, even though he lives in the
United States, he comes back every year with his daughter,
writes: "It is a unique place on earth where my child can experience a freedom of exploring a small town, independence
and beautiful walks in nature without the fear of anything
happening to her since everyone knows each other."

Sarov, formerly Arzamas-16, a city of ninety-two thousand,
which is still an important center for nuclear missile development, has also fought to restrict entry. It was disappeared
from the map in 1946 but remains closed off through local
determination rather than Moscow edict. A town tour guide,
Svetlana Rubtsova, explained to Russian journalists, "Being
part of a closed city gives you a feeling of comfort and protection—that people of this city are all together your family." Sarov, like a number of other restricted cities, is also an ethnic
Russian enclave, situated as it is in the ethnically mixed and
potentially separatist region of Mordovia. By remaining closed
"we defended it from chaos," says Sarov resident Dmitry Sladkov. An urban planner by training, Sladkov moved with his
family from Moscow in 1992 in order to escape the disorder
engulfing the capital.

In an era when claiming to be open to the world can seem
mandatory for cities that wish to prosper, the dogged survival
of closed places may appear shortsighted and misanthropic.
But Dmitry Sladkov's desire to flee with his family from the
"chaos" of open cities is not a uniquely Russian sentiment. It
isn't just in Russia that people are building closed communities. As modern cities around the world have become increasingly unpredictable and fragmented, people with enough

money have either moved out to villages, turning them into urban exclaves, or created gated, safe havens within the city. If we don't call Zheleznogorsk a closed city but a gated community it suddenly becomes not an echo from history but a very contemporary reflection of urban distrust and consumer choice.

But living in a gated community still has its problems. In Zheleznogorsk, besides being vetted by the security services before being allowed to visit, there isn't much for visitors to do. The Motherland movie theater in the center of town and one restaurant seem to be pretty much it. "It's difficult to start a business in a closed city," one local resident told the *Russian Gazette*. "The process requires many agreements, so there's no competition." For a fun night out she has to drive the forty miles to Krasnoyarsk.

How can it survive? Although its anchor industry, plutonium production, has been shut down, Zheleznogorsk has learned to reinvent itself in a number of ways, and there are plenty of other types of manufacturing that are attracted by complete privacy. Zheleznogorsk now nurtures a range of high-tech and "sensitive" forms of production. Three-quarters of Russia's satellites are produced in the city, including all of the GPS satellites. Israel, Indonesia, Ukraine, and Kazakhstan have all bought satellites made in Zheleznogorsk. Another niche that is opening up for the town is storing nuclear waste. An underground laboratory is being built that will investigate how much nuclear waste can be buried in the surrounding hills. It's the type of project that would be controversial elsewhere but that is facilitated in Zheleznogorsk by the pliant mindset of the locals, who have learned not to question those in authority: of its ninety thousand residents, only fifty bothered to look at the application details for the waste storage project.

Zheleznogorsk has successfully made the transition from a communist to a capitalist closed city. Its broad avenues may look like a Soviet stage set, but this is a place that says less about the past than about the high levels of privacy and security that are being demanded by contemporary companies and contemporary citizens.

The Underground Cities
of Cappadocia

38° 22' 25" N, 34° 44' 07" E

For a while the history of human dwellings seemed to be one of ascent: lifting us upward, plucking us out of dark caves, and placing us higher and higher aboveground. The dream house of modernity is the penthouse, not the pit. Of course such high-flying aspirations require a multitude of underground pipes and wires, but the rattling metros and oddly lit corridors and vents that service the sun-kissed surface dwellers have long been dismissed as the metropolis's idiot twin, useful but unlovable.

Today this stark divide is breaking down, for the lure of the subterranean is too strong. When nuclear apocalypse threatened, the only secure place seemed down there, and as we run out of space on the teeming topside, we are digging. Down there is where safety lies, in the one place we can escape the pollution and chaos of the scarred and scary surface and where we can control the temperature as the climate changes. New plans for underground cities are taking root from the Netherlands to China. The Amsterdam Underground Foundation claims that the public has now "embraced the mystical character of being underground." It's a provocative idea, because it suggests that we're not going downward merely as a technical fix or to avoid the bad weather. There is something else; something down there; something we are drawn to.

Underground cities are, understandably, hard to find on maps. Conventional maps are good at representing surface features but have trouble making visual sense of the multistory city. This is one of the reasons that buried places get overlooked and forgotten only to be rediscovered years later. The ancient underground cities that are still being uncovered in the Cappadocia region of eastern Turkey are a case in point. Some local experts believe that anywhere from thirty to two hundred may await discovery. Although that number is likely to include the ruins of cave monasteries and rock-cut villages, it is probable that we have seen only a small portion of the region's subterranean urban heritage. The largest underground city that we know of, Derinkuyu, came to light in 1965 and still remains only partially excavated. It was discovered by chance when a local resident was clearing out the back wall of his cave house. The wall gave way, revealing another chamber. And this chamber led to another, and then another. To date, Derinkuyu has revealed eight stories of underground rooms, and was large enough to have accommodated thirty thousand people. It had living quarters in its upper layers, as well as wine and oil presses, stables and food halls. A staircase leads from the third story down to cellars, storage rooms, and a church, carved out in the shape of a cross, at the lowest level. Derinkuyu also contains miles of man-made tunnels, one of which leads six miles south to another large underground town, called Kaymakli.

Buildings aboveground are designed around a supporting structure and take access to the air for granted. The underground cities of Cappadocia turn that model upside down, as excavators would have started by digging ventilation ducts and then worked outward from these shafts, opening up rooms and corridors, stables and sleeping quarters.

We find some clue as to why they were built in the place names: Derinkuyu means "deep well" in Turkish and used to be known as Malagobia, derived from the Greek for "difficult subsistence." In the first century B.C., the Roman architect Vitruvius proposed that the rock houses of Cappadocia were first built by the Phrygians some five hundred years earlier. Vitruvius wrote that because the Phrygians live in a "country destitute of timber," they "choose natural hillocks, which they pierce and hollow out for their accommodation." The hardened volcanic ash that covers the area, called tuff, is stable and easily carved. However, the progenitors of the local habit of cave-dwelling remain a source of controversy. The legendary German archaeologist Heinrich Schliemann sided with Vitruvius, but others push the origins of carved cities back further, to the Hittites, some one thousand years before the Phrygians. In any case, neither of these ancient peoples was directly responsible for either Derinkuyu or Kaymakli, for both were dug out from the eighth century A.D. onward by Christians. At that time Cappadocia was a lawless frontier region of the Byzantine Empire, and its long-established Christian population was suffering from periodic waves of invasion and banditry. In response to these threats, local Christians developed the region's existing architectural traditions to create underground settlements that were big enough to house whole communities.

Both Derinkuya and Kaymakli are built defensively. Entrances are small; each level can be sealed off with massive stone doors. Moreover, the spots where the numerous air ducts break the surface to the outside world are well disguised. There are also several cisterns and wells in the bottom levels, which suggests that inhabitants could lie low for extended periods. The question of how long people were down there at any one time hasn't been resolved. Most authorities suggest that we should look at the underground cities as boltholes, big enough to accommodate livestock and almost everything else that could be shifted, but occupied only when it was unsafe to go outside.

The troglodyte habit clearly bit deep, since the local population still lives by it. The outer and upper chambers of Derinkuyu, which spreads out underneath the surface town of the same name, have long been put to use as storage areas or stables. Many of ancient Derinkuyu's entrances are inside private houses, and people still draw water up from below through the city's old ventilation ducts. The tradition of underground construction is alive throughout Cappadocia, and the last century saw large subterranean stores being built across the province to house vegetable and fruit crops. A large percentage of Turkey's lemons and potatoes spend the spring and autumn in these cool caves.

These contemporary echoes speak of continuity, but the underground cities also have more uncomfortable messages.

They symbolize a buried history of religious plurality in what later became a purely Muslim nation. Over the last five hundred years the Christian population of Cappadocia dwindled away. The final remnants of a significant Christian presence in Turkey fled in the early twentieth century, victims of the 1923 ethnic population exchange between Turkey and Greece. Nearly all surface-level Christian villages have been erased from the map. One of the few places where the two-thousand-year history of distinct Christian townships in Turkey is preserved is under the ground in empty chambers.

Such places bring back memories of a past that has been pushed down but lingers on. Today tourists, Turks and foreigners alike, are drawn to these ancient Christian labyrinths in increasing numbers. The fact that they present a repressed stratum of Turkey's national story goes a small way to explain why. But the mystical character of the underground must also be acknowledged. As we descend into the dark we feel we are glimpsing something authentically archaic. That's the feeling, the prickle down the neck, the strange and discomforting urge to go deep. It's both a personal journey and a species-wide harking back to the precivilized, perhaps the prehuman. Michael Moorcock in *Mother London* and Peter Ackroyd in *London Under* make the case for the underground as a site of fear and desire, a dark font of possibilities. "All is true in underground writing," writes Ackroyd, and proceeds to speculate on a troglodyte race that has lived in London's tunnels since the Great Fire of 1666. As the underground cities of Cappadocia increasingly come to appear as ancient examples of a contemporary trend, and the subterranean becomes an ever more practical and occupied place, we will have to learn to talk about our expectations and fantasies of buried landscapes. As developers speculate about subterranean real estate, Cappadocia also serves as a reminder that only the truly fearful choose to live under the ground.

Fox Den

54° 58' 54" N, 1° 35' 21" W

Among our habitats other animals have created their own homes and pathways. Occasionally when I see an urban fox I have the urge to follow it, to know where it goes. There's a small fox that I've seen in the back lane behind my house, and I've been told it probably lives in the rhododendron and holly bushes at the overgrown western edge of Newcastle's Heaton Park. One day I ducked down and pushed my way among the branches. Some way ahead were the back fences of the big houses on Parkville. I came across some rubbish bags and a decaying wicker crib. The den was likely to be against the fence or dug from under one of these hefty shrubs.

In the countryside foxes like to spend the daytime resting in woods and fields. Only vixens retreat to their dens for any length of time, usually to give birth and look after cubs. They are not at all fussy about where they make their dens, the reused earths of badgers being a favorite site. But urban foxes have different habits. There are no badgers in the twenty-two acres of Heaton Park, which is made up mostly of woodland interspersed with playing fields, so its resident foxes have to dig out their own dens or make use of hollow trees or spaces underneath sheds or other debris.

All foxes like their den entrances to be in the sun. They also prefer light, well-drained soil and to have a number of emergency exits. All of which meant that if the hole in front of me was a fox den, it wasn't a very good one. There was a scrabbled, gloomy entrance about a foot across at the base of one of the trees, but I could see the faint light of water from inside the hole. I squatted and reached down; my hand dipped into a cold pool. This den had been flooded by the winter rain, and the animal must have moved on.

There are few things more thrilling than finding an animal's home, even when it is empty. Who doesn't want to pick up a fallen nest, to know its weight and peer inside and touch its bedded center? Even if our intent is to destroy them, the intimate care of the wasp's nest and the ant's miniature tunnels draws in our eye. It's not just the intricacy of these places that is absorbing but their indefatigable energy. My search for a fox's den in the hidden, marginal land of the park initially felt like a hunt for something fragile and rare, but if we didn't keep pouring on concrete and tarmac, all the places we call home would soon succumb to a quiet colonization. For a place-loving species, watching our place being overrun and turned into their place is a common fantasy. It's a possibility that is both appalling and fascinating. In *The Drowned World* J. G. Ballard argued that there is an atavistic desire for this kind of submergence lodged in the deepest, oldest parts of the human brain and based on a genetic memory of life's kindred emergence. Ballard's thesis could be mapped onto biologist Edward O. Wilson's notion that humans are programmed to be "biophiles" and have an evolutionarily determined love of living things. I would add that our dark fantasies of nature's revenge can also be seen as a byproduct of our suppression of nature. It's because we keep pushing other species down that the idea of their return haunts us.

We want to share the city even as we insist it is ours alone. The fact that the urban fox's habitat is spreading delights and alarms us. They have been living in British cities since the 1930s and have achieved a population density of up to five family groups per square kilometer. It was thought for many years that urban foxes were a uniquely British phenomenon. But in the 1970s they began appearing in cities as far apart as Oslo, Århus, Stuttgart, Toronto, and Sapporo in Japan. Wildlife ecologists who have been tracking them have found that rural and urban foxes have become markedly different, to the extent that one study discovered that "the border between the city and the surrounding grassland and forest was hardly ever crossed" by the two breeds. Another researcher discovered a "reduced gene flow between urban and rural populations." The urban fox also has a different diet and a different relationship to humans and its landscape. There have been at least two substantial projects on what ecologists call urban foxes' "daytime harborage." One, from 1977, was carried out in London and tracked down 378 foxes. Nearly 60 percent of them bedded themselves down in "gardens, sheds, cellars, houses"; the rest were found in sewage stations, in builders' yards, on vacant land, and in parks. Cemeteries and railway lines also proved popular. It is surprising that more people don't trip over them.

A second, more recent study carried out in Melbourne found a preference for "exotic weed infestations." The removal of these exotic weeds, the Australian experts concluded, will "assist in reducing the abundance of urban foxes." In Australia, foxes are often seen as a destructive alien species. This is the direction that a lot of academic ecology seems to gravitate toward: identifying habitats is all about finding better ways of stopping animals from spreading. In continental Europe the latest research on urban foxes has been driven by the discovery that they sometimes harbor a parasitic disease called alveolar echinococcosis, or small fox tapeworm. It's an infection that can be transmitted to humans, causing large cysts that require chemotherapy. Although the Institute of Parasitology at the University of Zurich has shown that distributing bait that expels the parasite is a more effective and sustainable way of dealing with this problem, the threat of the spread of small fox tapeworm will provide considerable ammunition to those who want to launch a war against the urban fox. However, the British, who have the longest experience of dealing with urban foxes, have come to the conclusion that trying to exterminate them is futile. Local authorities have given up that costly effort and found that the fox population soon reaches a state of equilibrium that foxes and most humans can live with.

I hastened away from the dark patch of wood, though retreating backward out of the bushes wasn't easy. Dirty water trickled off my hand and I felt a little cheated. Urban foxes may be fairly common, but they have developed the knack of disappearing into the city, so when they do appear it often takes people by surprise. That morning I'd read that construction workers in London had discovered a young fox living off their leftover sandwiches on the unfinished seventy-second floor of the Shard, the UK's tallest building. They decided to take the fox to an animal shelter, which released it back into the city. It was a gesture that carried a different type of ecological argument: while it's occasionally difficult to share the city with foxes, it is also inevitable. The release of the fox also acknowledged a bigger idea: that topophilia and biophilia are mutually sustaining—or, to put it another way, that accepting that the city is a multispecies environment benefits us all by enriching, enlivening, and, ironically, humanizing our sense of place.

North Cemetery, Manila

14° 37′ 53″ N, 120° 59′ 20″ E

Who is more off the map, the living or the dead? Most of our streets, towns, buildings, and nations are the creations of the dead and carry their names. The living parade in their kingdom like ghosts. In an era that mythologizes the living as go-getting gods, able to reshape and revolutionize everything we touch, it is an uncomfortable situation. It's this mismatch, this discomfort, between our self-regard and our sneaking sense of watery insubstantiality, that explains much about our horror of the dead. We resent the power that the dead have over us, their effortless capacity to reduce us to shadows.

One way to rid ourselves of this tomb envy is to come to an accommodation with the deceased. We'll stop being frightened of them if they give us access to their places of rest. The living have a lot to gain from the arrangement, since it could result in a lot less fear and a lot more housing space. This brings me to North Cemetery in Manila, the densely packed megacity that is the capital of the Philippines. North Cemetery represents a new kind of urban environment, the lived-in graveyard, and has between three thousand and six thousand living residents, many of whom live in and around its substantial family tombs. High city rents make the free space of the cemetery an attractive option for the poor, but this is more than just another story about destitution. It's also about a realignment of people's spatial relationship with the dead.

With the growth of the world's population and the mounting challenges of making a living from the countryside, cities around the world have been getting bigger and fuller. As demand has increased, rental prices have become too expensive for many ordinary people. Cemetery living is one of the solutions to this problem. It's not as visible in the United States or Europe but it still happens. I used to have a colleague who lived for years in a camper van in a graveyard in northern England. He found the right patch of ground and got by on very little. But his housing choice remained an eccentric one.

You have to go east to find whole communities living in graveyards. There are numerous reports of cemetery living from India, Pakistan, Chechnya, and more recently in Libya, where nearly two hundred families have moved into the Al-Ghuraba cemetery in Tripoli. Like many cemetery dwellers they are very poor, with nowhere else to go. But the best-known and certainly the biggest living cemetery, Cairo's City of the Dead, proves that these places can be far more than repositories of the desperate. Given time, they can establish themselves as thriving and diverse economies. In the five cemeteries that make up the City of the Dead, about fifty thousand live within tombs and another half million in houses put up between tombs. If this were only a field of slab headstones, no such community would have arisen. In the West substantial tombs are a rarity, being the preserve of wealthy dynasties; the rest of us get little more than a boot scraper. But Egyptian cemeteries were never designed only for the dead. Traditionally in Egypt it was expected that mourners—a role given to female relatives—would live with the deceased for forty days. So the family tomb was constructed as a complex, with additional rooms and a courtyard. Egypt also has an ancient tradition of seeing cemeteries as places where the living and dead come together. In fact, the City of the Dead is better seen as just another urban district. It has its own shops, schools, and a clinic with a maternity wing, as well as electricity and running water. Since the City of the Dead began to be permanently occupied in the 1950s, several generations have been born there, often sharing the same tomb with their parents and grandparents. The Italian anthropologist Anna Tozzi di Marco, who has studied and lived in the City, refutes the idea that it is a place of desperation. Instead, she offers a portrait of a place with its own class structure, a city within a city, in which people can make a success of urban living, rent-free.

The City of the Dead is a fully formed inner-city suburb. By comparison, North Cemetery in Manila is smaller and more specialized. Like the City of the Dead, it too started to be occupied from the 1950s and also offers living spaces in tombs, although they are not nearly as palatial as some of those in Cairo. Having grown for sixty years, the Manila cemetery also has its own neighborhoods, some of which have long been just as self-sufficient as, and certainly safer than, the slums outside. It also has amenities, such as several mini-markets, a restaurant, and sports facilities. Electricity is illegally cabled in from beyond the cemetery. However, while the City of the Dead seems to fit into and grow out

of Egyptian culture, North Cemetery is a far more pioneer-
ing, transgressive place. Catholic Manila has no equivalent of
Cairo's Islamic and pre-Islamic traditions of extended live-in
mourning, and as a result the residents here see themselves
as overstayers and as out of place. They go to great lengths to
make themselves useful to the everyday life of the cemetery,
caretaking family tombs and undertaking tasks such as carrying coffins and sealing up crypts. They get out of the way and
live somewhere else on November 1 and 2, the Days of the
Dead, when many Filipinos come and visit their ancestors.
Behaving like a closed community of guardians, the cemetery
residents have worked out a respectful if rather nervous relationship with both the living and the dead.

One resident, Bobby Jimenez, explained to a roving journalist, Kit Gillet, "We do occasionally go outside the walls—
to walk the streets—but mostly we stay inside." He went on
to describe the precarious nature of life in North Cemetery:
"Sometimes we have police raids, so it is important to try to
get approval of the family owners of the tombs. If you have
a piece of paper or a deed from the family saying you have
a right to stay there it is OK." Even someone like Clare Ven-
tura, who was born in the cemetery and brought up her three
children within its walls, said, "I've had to teach myself to like
living here." As she explained to an interviewer on National
Public Radio, "This is where I have a chance to earn a little
bit. You get used to it, and it's a lot safer here than most places
outside." Other tomb dwellers, like Boyet Zapata, complain
that the restless spirits of the newly dead can interfere with
their lives, taking over their bodies.

But the residents have reached an understanding with the
departed, based on reciprocal respect and care: they look af-

ter the dead's resting places, and in return the spirits of the dead, for the most part, leave them be. Over recent years it is not the dead but an influx of squatters, including alcoholics and drug addicts, that is disturbing residents' peace of mind in North Cemetery. These interlopers misuse the tombs, hassle mourners for money, and disrupt burials. Their behavior shatters the special kind of patience that is required to make North Cemetery a place that works for both the living and the dead. This influx of undesirables has also stung the city authorities into threatening to clear out the entire community. To be swept out into the hostile world would be a terrible injustice to the people who, for generations, have cared for this place, but even if this happens, they and others like them would soon find a way back in. With city rents staying stubbornly high, for many people coming to an accommodation with the dead is one of the few viable ways to sustain life in the city.

North Sentinel Island

11° 33' 20" N, 92° 14' 77" E

Wild men, estimate more than 50, carrying various home-made weapons are making two or three wooden boats. Worrying they will board us at sunset. All crew members' lives not guaranteed.

This radio distress call was received at the Regent Shipping Company in Hong Kong on August 5, 1981. It was from the captain of the MV *Primrose,* a cargo ship heading to Australia through the Bay of Bengal. The ship had struck a coral reef and was grounded some one hundred meters away from dense forest. The *Primrose* had run onto the shore of the only island in the world entirely occupied by an "uncontacted" indigenous people. There are roughly a hundred inhabitants, and their language, religion, and customs remain unknown. The outside world calls them the Sentinelese, and they live on a round isle that is five miles across, North Sentinel, one of the necklace of 361 islands that make up the Andaman and Nicobar Islands, a Union Territory of India that lies about eight hundred miles to the east of India.

The captain had good reason to worry. The usual response of the Sentinelese to intruders is a hail of arrows. But this time the seas were rough enough to keep their canoes at bay, and their unfletched arrows, which have a range of only forty meters, fell into the water. A long week passed before the crew of thirty-three were lifted off their vessel by civilian helicopters.

DNA results from related tribal groups in the Andamans indicate that the ancestors of the Sentinelese migrated to the islands from Africa some sixty thousand years ago. North Sentinel is the last redoubt of an ancient community. The island has no natural harbors and is surrounded by reefs and year-round rough seas. It is a fortress against the world. For many years all attempts to approach the island were met with the same level of hostility. In 1974 a film director had received an arrow in his thigh while laying out a selection of ingratiating gifts on the beach: pots, pans, a live piglet, and toys. However, in the 1980s and early 1990s the Indian authorities began a concerted effort to win the islanders over. Anthropologists and local officials made regular forays to the shoreline, always bearing gifts. After numerous hesitant and unsuccessful efforts came the breakthrough. On January 8, 1991, the front page of the Andaman newspaper, the *Daily Telegrams,* declared FIRST FRIENDLY CONTACT WITH SENTINELESE. The story told how, having left the usual offering on the shore—in this case a bag of coconuts—and retreated to their motor launch, officials had witnessed the Sentinelese coming out of the forest to collect them. The sensation was that this time they came unarmed. In the afternoon the Indians went back and found more than two dozen native people waiting for them. The visitors observed a small but telling incident: a young woman walked over to a young man who was pointing his loaded bow at the strangers and with her hand she pushed his arrow down. The man then buried the weapon in the sand. Pleased with the way things were going, one of the officials, the director of tribal welfare, decided to celebrate by throwing numerous coconuts to the assembled crowd. These seem to have been well received. The only person who can claim even superficial knowledge of the Sentinelese, the Indian anthropologist T. N. Pandit, explained to a reporter in 1993, "They may not have chiefs but a decision had obviously been taken by the Sentinelese to be friendly towards us. We still don't know how or why."

The nascent relationship didn't last long. In 1996 trips to North Sentinel came to a stop, and since then the islanders have been left alone. The Indian policy of noncontact was bolstered the following year by a bad experience with another previously uncontacted and once hostile tribe, the Jarawa, on the island of Main Andaman. The "Jarawa crisis" started in late 1997 when, having been encouraged to come out of their forest seclusion, the Jarawa upset local mores by wandering naked into villages and taking whatever goods they wished. They also became prey to sexual exploitation and measles, a disease that today threatens the tribe with extinction. Contact with the Jarawa had created a headache that the authorities didn't want again.

Stephen Corry, the director of the indigenous rights charity Survival, estimated in 2007 that there are 107 "largely uncontacted tribes in the world." But, he adds, "they remain separate because they choose to, and with good reason." Although hidden tribes are usually associated with the Amazon jungles, nearly half of the world's known uncontacted groups are in West Papua, on the island of New Guinea. Many of these people are in flight or in hiding from the army and settlers from Indonesia, which treats West Papua like a colonial fiefdom. The result of contact for these hidden peoples would, at the very best, be cultural decay, but another probable outcome would be death through disease and assault. Another likely fate is for these peoples to become the object of touristic curiosity. "First contact for cash" is one of the holiday options adventuresome tourists in West Papua can opt for. Clients are taken deep into the jungle and often allowed to freely mix with "uncontacted" and "absolutely primitive cultures." A BBC interview with a "first contact" expedition leader in 2006 allowed him to make his pitch and to provide an ethical justification. Anyone and everyone, he says, should "have the right to see these kind of people." Given the tragic history of such contact, it's a grotesque kind of right. No one knows if the Sentinelese are aware of other indigenous peoples of the Andaman Islands or what has happened to them. Today such groups make up no more than 1 percent of the population. The names of the many dead tribes form a litany of lost sounds: Aka-Bea, Akar-Bale, A-Pucikwar, Aka-Kol, Aka-Kede, Oko-Juwoi, Aka-Jeru, Aka-Kora, Aka-Cari, Aka-Bo.

The fact that the Sentinelese don't care for strangers was driven home again in 2006 when they killed two fishermen. Sunder Raj and Pandit Tiwari had anchored offshore but during the night their open-topped boat drifted onto the beach. In the early morning other fishermen shouted to them, to try to wake them, but got no response—reports later suggested that they were "probably drunk." It was also rumored that later that day the two fishermen had been eaten, but when an Indian coast guard helicopter hovered over the beach, a more mundane reality was revealed. The draft from the rotor blades exposed the bodies of the men buried in shallow graves. As this discovery was made, the Sentinelese were shooting arrows up at the helicopter, so no attempt was made to recover the corpses. The Andaman Islands police chief later claimed that "once these tribals move to the island's other end we will sneak in and bring back the bodies." But to date they remain on North Sentinel.

Should the murder of these two innocent men go unpunished? Why is it so obvious, as it was to the two men's relatives, that prosecuting the tribe members wouldn't be an act of justice? North Sentinel isn't part of our modern world and it doesn't ask anything of us except to leave it alone. An officious-sounding government document, the *Master Plan 1991–2021 for Welfare of Primitive Tribes of Andaman and Nicobar Islands,* published in the Andaman capital, Port Blair, in 1990, turns out to make a lot of sense. It concludes that "the Sentinelese do not require the benevolence of the modern civilization," adding that "if at all they require anything, it is noninterference." The *Master Plan* proposes a policy of stay clear. "What right does modern man have to interfere in the totally isolated tribal life of the Sentinelese? What right has he got to decide unilaterally to impose his 'friendship' on the Sentinelese who have been vehemently resisting it?"

So no more gifts are allowed, no more toys and coconuts, only an occasional observation from what the *Plan* calls a "respectable distance, say 50 meters from the shore." For the past fifteen years or so this approach has been enforced. No one is allowed near. One day almost everything the Sentinelese know and value will disappear, as it has for every other once uncontacted community. But for the time being North Sentinel Island is theirs and theirs alone.

NO MAN'S
LANDS

Between Border Posts
(Guinea and Senegal)

12° 40' 26" N, 13° 33' 32" W (border point)

"No man's land" is a term that, to the modern ear, can sound like stepping onto a battlefield. In fact, the phrase refers back to the idea of unclaimed land (recorded as "names-maneslande" in the Domesday survey of England of 1086) and still carries an echo of perennial hopes for free land, for places beyond the control of others. Ordinary places become extraordinary in no man's land. Such in-between places remind us how dependent we are on borders—that our sense of order and certainty draws deeply from the knowledge that we are in governed territory. No man's lands may be vast stretches of unclaimed land or tiny scraps left over from the planning of cities, though the uncertainty of the no man's land is especially keenly felt in places that the outside world refuses to recognize or that appear to be between borders. The notion that places might slip down between borders led me on a geographical quest. I went looking for the farthest possible distance between the border posts of two contiguous nations, to see how far they could be stretched apart.

Most border posts face each other. A change of signage, a different flag, a line on the road, all combine to signal that no sooner have you stepped out of one country than you have arrived in another. But what happens if you keep on opening up that space? A few years ago, with the help of hours spent blinking at the tiny fonts favored on travelers' Internet forums, I found what I was looking for. Along a road between Senegal and Guinea in West Africa the distance between border posts is 27 kilometers. It is not the world's only attenuated border area. The Sani Pass, which runs up to the mountainous kingdom of Lesotho from South Africa, is the most famous. It's a rough road, although much visited by tourists in 4 x 4s seeking out the highest pub in Africa, which sits near the top of the pass. The drama of the trip is heightened by the thrill that comes from learning that this is no man's land. The South Africa border control, complete with "Welcome to South Africa" signs, is 5.6 kilometers away from the Lesotho border office. Another specimen is to be found in the mountainous zone between border posts on the Torugart Pass that connects China and Kyrgyzstan. Central America also has a nice example in Paso Canoas, a town that can appear to be between Panama and Costa Rica. It is habitually described as no man's land because, having left through one border post, you can go into the town without passing through immigration to enter the other country. Some visitors relish the impression that the town around them is beyond borders. Partly as a result, Paso Canoas has developed a darkly carnival atmosphere, as if it were some kind of escaped or twilight place.

What these gaps reflect back at us is our own desires, especially the wish to step outside, if only for a short time, the claustrophobic grid of nations. We probably already suspect that it's an illusion. Shuffling forward in a queue and making it past the passport officer does not mean you are, at that exact moment, leaving or entering a country. Such points of control exist to verify that you are *allowed* to enter or leave. Their proximity to the borderline is a legal irrelevance. Yet this legal interpretation fails to grasp either the symbolic importance of the border point or the pent-up urge to enter ungoverned territory. The fact that Paso Canoas is split by the Panama–Costa Rica border rather than actually being between borders doesn't stop people from describing it as an "escaped zone." Similarly, the steep valley up the Sani Pass is nearly all in South Africa, and the road down from Senegal into Guinea is always in one nation or another, but that isn't how travelers experience it or even what they want.

The attraction of these in-between spaces has a lot to do with the fact that they are on land. Going through passport control at an airport provides no comparable thrill, even though international airspace is far more like a genuine no man's land than any number of dusty miles on the ground. It seems that escaping the nation-state isn't all that is going on here. There is a primal attraction to entering somewhere real, a place that can be walked on, gotten lost in, even built on, and that appears to be utterly unclaimed.

Some of the overland tourist trips that occasionally rumble along the Senegal–Guinea highway offer camping in the no man's land as part of the package. Like other examples, it's a zone that provokes people to muse on allegiance and belonging. In his essay "Life Between Two Nations," the American travel writer Matt Brown describes encounters with villagers along the Senegal–Guinea road that provoke speculation on the nature of national identity:

I stopped my bike to chat with the woman pounding leaves. I asked in French (my Pular only goes so far), "Is this Guinea?"

"Yes," she answered.

Surprised that she even understood French, I posed a follow-up question. "Is this Senegal?" I asked.

"Yes," came the reply.

A little later Brown sits on "a nationless rock" and imagines these villagers as freed from the "archaic, nonsensical national borders drawn up by greedy European leaders at the Conference of Berlin over 100 years ago." Stretching out border posts does seem to break the seal on the national unit. The resultant gap may not be of much legal import, but for travelers on the ground it creates a sense of openness and possibility.

Yet while travelers may relish this expansiveness, the consequences for those who have to live and work in such places can be less positive, such as heightened insecurity and a sense of abandonment. This is one of the reasons why African states have been trying to close the gap in such anomalous spaces. The African Development Fund, which supports economic infrastructure projects across the continent, has made "establishing juxtaposed checkpoints at the borders" of its member states a priority, including at the Guinea–Senegal border. What most concerns the fund's members is the impact that these distant border posts have on the flow of trade. Along the Guinea–Senegal route there are nightmare tales of vehicles being sent back and forth by officials who keep asking for new documentation or demanding new bribes. In-between land can easily turn into a place of bureaucratic limbo where both travelers and locals are uniquely vulnerable to tiresome and corrupt officialdom. Patches of ground "between" nations are places that can be thought of as free, but they are also places where we are reminded why people willingly give up freedoms for the order and security of being behind a border.

Bir Tawil

21° 52' N, 33° 41' E

It seems incredible that anywhere could be so ill thought of that no one wants it. Bir Tawil, a 795-square-mile trapezoid of rocky desert between Sudan and Egypt, is such a place. It is not just a no man's land; it is actively spurned. It also appears to be the only place on the planet that is both habitable and unclaimed.

The Bir Tawil anomaly opens up a new perspective on world history. It is the history of the struggle *not* to occupy territory, and sounds like history written back to front. Asserting ownership over place is at the root of many of the world's enmities and identities. It's no surprise that we tend to assume nations want to continually grow; that the border, much like the fence put up by an inconsiderate neighbor, is always being pushed to the maximum extent. But Bir Tawil reminds us that nations are defined by their limits: that land is not always wanted, and that for every claim on a piece of ground there must be many acts of denial and avoidance.

For Sudan and Egypt the point of not wanting this land-locked region is that it bolsters a claim to an even bigger and

more useful parcel, the 8,000 square miles of the Hala'ib Triangle, which faces the Red Sea. Their dispute arises from the
existence of two different versions of the border that separates Egypt and what used to be called Anglo-Egyptian Sudan. Both were drawn by the area's British administrators.
The first is from 1899 and is a 770-mile-long straight line
across the desert, and is the border Egypt is keen to keep. It
gives Bir Tawil to Sudan but holds the valuable Hala'ib Triangle on its side. The Sudanese don't accept this border and
point to another, drawn in 1902, which is mostly straight but
toward the coast begins to change course, giving a tongue of
land along the Nile to Sudan (called the Wadi Halfa Salient)
but also the Hala'ib Triangle. The 1902 map gave both places
to Sudan because they were considered by the British to be
ethnically and geographically linked to the south. The mak-
ers of the 1902 map applied the same logic to scoop the border
southward and place Bir Tawil in Egypt. They considered Bir
Tawil to belong ethnically in the north because it was used
for grazing by the Ababda, a nomadic tribe that lives in southern Egypt.

For many decades the 1902 borders were not seriously disputed. In the early 1990s, however, Sudan granted oil exploration permits for the Hala'ib Triangle. Egypt responded by
occupying the area and claiming its right to defend the 1899
border. In response, the Sudanese have offered gestures of de-
fiance. In 2010 a government official tried to get into Hala'ib,
with the apparent idea of getting the locals to vote in Sudanese elections. If the plan had worked, it might have backed
up Sudan's claim, but the official wasn't allowed in, and for
now the Egyptians seem to have succeeded in claiming the
Hala'ib Triangle and disclaiming Bir Tawil.

In the meantime, Bir Tawil has become ever more un-wanted. Although the name means "tall well," a prolonged drought has removed what little agricultural value Bir Tawil ever had. Satellite images appear to show that across its bar-ren miles there is not a single building. Even its desert tracks are now disused, disappearing reminders that this was once Ababda territory. The Ababda took little notice of the region's national borders and had their own distinctive ethnic heri-tage. A 1923 issue of the *Journal of the Royal Anthropological Institute of Great Britain* has photographs of Ababda men with thick, tightly braided hair alongside local myths about them, for they were seen as mysterious and ancient even by other desert tribes. It was said that "when followed up into the des-ert" an Ababda "vanishes from sight after going 200 or 300 yards." Moreover that their glance is very dangerous to others and "they can bring moving objects to a standstill, when at a considerable distance from them."

The Ababda have moved away, but an important part of their story remains rooted in this place. Thus, to say that Bir Tawil is unoccupied is not to say that it has no history or that it's anyone's to take. It is a point worth making since the un-claimed space of Bir Tawil has become a favorite fantasy item among the Internet's would-be nation builders. Indeed, these days real information about the place is obscured by myriad websites and online disputes between fictional kings, emirs, and presidents of Bir Tawil.

These playful claims misjudge Bir Tawil's somber reality, but it is hard to be too disapproving. Bir Tawil excites the geo-graphical imagination because it disorients our expectations of the modern world, and more specifically our expectations of what nations and borders are trying to achieve.

It seems natural to define the world around what is sought
after, but geopolitics can also be considered in terms of what
is not wanted. There are a number of ways this happens.
First, as we have seen with Bir Tawil, there are "anti-claims"
designed to bolster positive claims. Such cases are not uncommon, although usually there is a keen recipient for such
apparently unloved places. The borders of China have a number of them. A recent summary of the situation showed that
China has given ground in 17 out of 23 of its ongoing border disputes, giving up 1.3 million square miles of land. In
the long-running dispute between Greece and Turkey, whose
peoples were once intermingled but, over the course of the
last century, became isolated into separate, ethnically discrete
states, a lot of attention has been given to defining where is
not "historically" Greek or *not* "historically" Turkish. Greek
and Turkish irredentism is shot through with as many denials
as affirmations. In the quid pro quo of territorial disputes, declaring a lack of interest in a region often turns out to be the
key claim.

Bir Tawil is easily overlooked on the world map as an oddity, an area of minor confusion where geopolitical certainty
has broken down into a series of dashed lines. Yet its story is
of universal importance. For Bir Tawil is one of the few places
on earth where one of the key paradoxes of border-making is
being explicitly played out. Borders are about claims to land,
but as soon as you draw one you limit yourself. Every border is also an act of denial, an acknowledgment of another's
rights. By contrast, the claim to want no borders, much prized
by corporate executives and anticapitalist activists alike, is a
claim to the whole world. Borders have a far more ambivalent

and complex relationship to territory; they combine both arrogance and modesty, both demand and denial.

Nahuaterique

14° 03' 05" N, 88° 08' 57" W

When borders change, some unlucky communities end up on the wrong side of the wire and wake up to find they are foreigners in their own country. This has been the experience of the people of the remote mountain region of Nahuaterique, which was handed over by El Salvador to Honduras in 1992. The story of Nahuaterique also shows how places once thought to be so important that they were worth fighting for are often forgotten once the battle is over.

Honduras and El Salvador have been bad neighbors for a hundred and fifty years. They have been fighting over their shared border for much of this time. The most recent conflict was a four-day war in 1969. It's often referred to as the Soccer War since it was preceded by clashes between fans of the two national sides during the games they played in the second North American qualifying round for the World Cup. The war, however, wasn't actually caused by a soccer brawl. Its true cause was demographic pressure. For years there had been a steady influx of landless people from the small and crowded country of El Salvador into Honduras, which is four times as big. They were looking for work and land to farm and moving across an ill-defined and disputed border. In retrospect it looks as if they were migrating from one country to another, but that isn't how many of them understood it at the time. As far as they were concerned, they were just moving east, going up into the relatively empty lands of the mountains.

The Soccer War has another, more fitting name: the War of the Disposed. However innocent these settlers' intent, they were treated as illegal immigrants and as land snatchers. Thousands of Salvadorans were thrown out and new laws were introduced that took land away from Salvadorans in Honduras and gave it to native Hondurans. It was this bitter intervention that started the war. It lasted only four days because the Salvadoran army quickly made deep inroads into Honduras until, under considerable international pressure, it was forced to fall back.

A long series of border negotiations began, eventually ending up at the International Court of Justice in The Hague, and it wasn't until 1992 that a new and definite border was announced. Most people seemed content with the outcome. Roberto Hidalgo Castrillo, the Salvadoran ambassador to the Netherlands, announced that "we can celebrate with great joy." El Salvador lost Nahuaterique, but this was just one of six disputed areas. The fact that a few small communities would soon find themselves on the other side of the border was seen as a price well worth paying. Twelve thousand Salvadorans found themselves in Honduras, while three thousand Hondurans were informed they were living in El Salvador.

The Salvadorans argue that the people of the twenty-one villages that are scattered across Nahuaterique have since been neglected by their feckless new owner. In fact, the Honduran government is not unsympathetic to the region's plight. Honduras is understandably concerned about having a lawless no man's land on its border, but its expressions of concern have been low-key. The people of Nahuaterique rarely vote in Honduran elections and have few real friends in its capital city, Tegucigalpa.

Writing for the Salvadoran newspaper *La Prensa Grafica* in April 2013, Siegfried Ramirez recalls how the villagers woke to find themselves abandoned. "When the people first heard the rumor that the land where they lived was not part of El Salvador but in Honduras," Ramirez writes, "many believed it was just a bad joke." Honduran officials soon arrived in the area demanding that its residents register their land as Honduran, but the registration process was never completed. As a consequence, only a minority of land in the area is legally owned. It has also taken decades for citizenship cards to be distributed, preventing residents from having any access to basic state services or getting a driver's license.

The villagers feel forsaken: no longer in El Salvador but disowned by Honduras. The public and legal services available in Nahuaterique remain nonexistent or unpopular. Although schools have been built, until recently many children preferred to walk three hours across the border to go to school in El Salvador. The 1992 border also put an end to the area's main trade, which was supplying timber to El Salvador. Overnight it turned from a legitimate business into illegal trafficking. The status of a uniformed state presence in the area is even more uncertain. The only sign of Honduran authority is a military post at Palo Blanco, and the soldiers there report that their only task is to secure the border from lumber smuggling. Locals complain that schoolchildren have been robbed by thieves right in front of the Palo Blanco post. When asked why they didn't intervene, the soldiers claimed that "school safety" was not their concern and, in any case, they didn't have the power to make arrests.

So at the moment one man, a wiry farmer called Marcos Argueta, is the law. *La Prensa Grafica* reports that as "he walks down the center of town, people look at him with respect." Argueta's authority is based entirely on a local and unofficial election, but it has thrust him into the limelight as the voice of the people. "Many people here didn't want to be Honduran," he explains, "but they couldn't leave as they didn't have land elsewhere." Since the transfer, Argueta says, "there have been serious issues with security. Anyone can come in — drug traffickers, criminals." It's a situation that he struggles to manage. Drunks and bandits are dealt with by Argueta through a combination of rough justice and wishful thinking. Along "with other reliable men" his technique for dealing with bad behavior is to knock the offending party to the ground and bind his hands and feet till he promises to behave or, ideally, leave.

Such is the despair in Nahuaterique that twelve of its residents started an "indefinite" hunger strike in 2012 outside the National Congress in the Honduran capital. In their press statement they demanded their own regional government as well as "schools, health centers equipped with personnel and medicines, agricultural support and the free movement of people and goods." It's a long list, but it does not meet with animosity in Honduras, where the public and political attitude mixes sympathy and neglect. Though the abandonment of Nahuaterique is widely reported in Honduras, there is no sense of urgency. It is pointed out that since 1998 dual citizenship has been granted to people in these once disputed borderlands and that they have the privilege of going freely between the two countries. The deputy minister of the interior and population, Salome Castellanos, has warned the villagers to stop complaining. What they need to do, he says, is to learn to live as Hondurans.

While the Honduran newspapers report on the plight of the people of Nahuaterique, they spin it into a good-news story of steady progress. Thus recent headlines in *Hondudiario* and *El Heraldo* have announced how thankful the people of the area are to the Honduran human rights ombudsman for sticking up for them and how much they are looking forward to working hard to make what one resident quoted in the papers apparently described as "a new Nahuaterique completely and actively integrated with the social, economic and political life of our new homeland, Honduras." Alongside these happy thoughts we find government promises of a police station to be built "in the coming months" and the appointment of a doctor for the area.

But if and when Nahuaterique does become "completely and actively integrated" into its new homeland, questions will still remain: Why did it take so long? After having fought so hard and having won back such a sizable chunk of territory, why did Honduras choose to turn its back on the region? It seems that, for Hondurans, the meaning of the place was all in the fight. It also suggests that Nahuaterique just has too many Salvadorans to ever be taken seriously in Honduran politics.

Twayil Abu Jarwal

31° 19' 2" N, 34° 48' 2" E

A place is not a thing, like a pencil or a watering can, something that can be thoughtlessly disposed of and replaced. The ferocity and ingenuity with which people hang on to the place they care about shows that it is a defining feature of who they are; that to lose one's place can seem like losing everything.

In an era in which the importance of place is often overlooked, we need to turn to desperate places such as Twayil Abu Jarwal, a Bedouin village in Israel's Negev Desert, to be reminded of the urgent and necessary nature of topophilia. It's up a dirt track off the smooth tarmac of Highway 40 that runs north of Beersheba. There are no road signs to the village and it doesn't appear on any maps. But like the forty other "unrecognized" Bedouin villages in the Negev, Twayil Abu Jarwal grips onto this bone-dry landscape with a stubborn energy. It has been demolished by the Israeli authorities so many times that accounts vary widely on the exact number, but conservative estimates suggest that the bulldozers have rolled in and pushed over some part of the village between twenty-five and fifty times. Today there are no permanent structures for its 450 residents, only tents and tin shacks.

In the aftermath of each demolition a group of villagers gathers to assess the damage. Israeli activist Yeela Raanan recorded one such exchange: "The bulldozer driver took his time," says one, "he worked slowly and thoroughly, he left nothing standing, nothing." This time, says another, "they buried alive the doves' hatchlings." But as soon as the bulldozers leave, the village reemerges—shelters and pathways are reestablished; it becomes a place again, awaiting the next visit from the Israeli authorities. Talking to an observer from Human Rights Watch, a village woman, Aliya al-Talalqah, describes how it takes "five to six days after the demolition to make these tents." In the meantime, everyone has to sleep "outside on mats, like wild animals, with the sun in the day and the cold at night, with small children."

Interviewed in the *Jerusalem Report,* Ilan Yeshurun, a local director of the Israel Land Authority, explained the ceaseless round of demolition by simply stating, "This is not a village." Without irony, he added, "It doesn't exist on any map or in any legal registration. It's only a village in the eyes of the Bedouin." In other words, it is because Twayil Abu Jarwal doesn't exist that it both can and has to be bulldozed again and again.

The tents of Twayil Abu Jarwal, flapping bleakly amid the rubble, offer a broken echo of a time when the Negev Bedouin were a nomadic people, moving with their sheep and goats across the desert. The Bedouin were pretty much ignored by the area's past rulers, the Ottomans and later, briefly, the British. Israel's was the first government that took an interest in them. From the 1960s the Israeli government pursued a policy of "sedentarization" and concentration, relocating the Bedouin in seven new towns in a triangle of land in the Negev called the Siyag. It was hoped that this ancient people would be reshaped into a modern community. In 1963 Israeli general Moshe Dayan looked forward to a time when the Bedouin would constitute "an urban proletariat" and each man "would become an urban person who comes home in the afternoon and puts his slippers on." But the Bedouin carried with them a huge sense of loss into the Siyag's new towns, and they were ill prepared for urban life. The new towns soon became associated with social breakdown, crime, and unemployment. Many drifted back to their ancestral land. No longer nomadic, and with an increasingly fragile connection to traditional Bedouin identity, the al-Talalqah clan chose to build the village of Twayil Abu Jarwal near their old tribal cemetery.

Settlements like Twayil Abu Jarwal deliver the goal of a sedentary lifestyle but on the Bedouin's own terms. The Bedouin sense of place, which once extended across the Negev Desert, has become anchored in such locales. But their attempt to make this transition on their own terms has been continually challenged. The urban planner Steve Graham offers the word "urbicide" to describe the Israeli government's policy toward the Palestinians, referring to the attempt to smash political resistance by breaking up the physical and social infrastructures of urban life. But at least the Palestinians have places to break. The problem for the Bedouin is that their villages are not even acknowledged. Ironically, as a cultural and ethnic group, the Bedouin receive a lot of attention. Their traditional clothes, food, and other items of ethnographic interest get tourist and state attention and even respect, but without the acknowledgment of place, respect for mere artifacts means little.

Why do we have such a hard time grasping why people care so much about place, even if it is only a few rubbish-strewn meters of scrub? It is a difficulty that is in part rooted in the nongeographic way we approach the task of acknowledging or recognizing others. The German philosopher Hegel argued that people need recognition from others in order to

achieve a sense of self. Hegel went on to claim that consciousness is always trying to make itself more pure and less dependent on dumb materiality. Thanks in part to such inter-
ventions, our ideas about what human liberation means have
become ever more untethered from the earth, drifting off
into abstract realms and leaving geography to become nothing more than a tedious list of facts. An obsession with the
struggle for free consciousness, filtered through Karl Marx,
framed the worldview of the last century's anticolonial intellectuals. Jean-Paul Sartre, Albert Memmi, and Frantz Fanon
turned the colonized world into a field hospital of psychoso-
cial trauma. Today the pain and humiliation of subject peoples has been fashioned into a series of sub-Hegelian clichés
about respect for "others" and respect for "difference." But all
this attention to the internal life of victims has obscured as
much as it has revealed, turning place into an irrelevance or
something meaningless and inert.

Place is the fabric of our lives; memory and identity are
stitched through it. Without having somewhere of one's own,
a place that is home, freedom is an empty word. Twayil Abu
Jarwal is just one ruined village hoping for recognition, but its
story—like the stories of the Negev's other unrecognized villages—is not only a local one. It reminds us of the necessity
of place and the battles that are being fought between those
who want to recognize places and those who wish to deny
them.

The grievance of the Bedouin is sharpened by the fact
that, while their illegal villages fail to make the map, illegal
Jewish farmsteads are being tolerated across the Siyag and
the wider region. Fifty such farmsteads have sprung up, many
with unapproved buildings. Rather than knocking them down,

the state facilitated their growth. One way this has been done was by extending Highway 6 into the Negev. In drawing the route of the new road, the country's transportation planners simply ignored the existence of the Bedouin's unrecognized villages. The road's blueprints show that it plows straight through a number of them. The highway will soon be on the map, but the villages beneath and around will remain invisible. Although recognition has been won for half a dozen Bedouin villages, the Israeli state's plans for the region involve forced resettlements and demolitions for the remainder. It seems likely that Twayil Abu Jarwal will be destroyed many more times before its inhabitants or the government gives up.

So what do you do if demolition is inevitable? The Bedouin are in a bind. If they challenge a demolition order, they have to admit that they built illegally and thus confess to a criminal act. With nowhere else to go, their claim on places like Twayil Abu Jarwal takes a perverse route, self-demolition. It is a final statement of control, a last claim to place. As one resident in the nearby unrecognized village of Wadi al-Ne'am explained to Human Rights Watch, "When they came there were three houses they wanted to demolish. We said, we don't want you to cause panic in the community so we'll demolish them ourselves. They were waiting outside the village, and we demolished them ourselves, and then they came back to check. We got tired of the threats, and that's when we decided to do it ourselves."

Even this act of self-destruction goes unrecorded. For these places never existed: the history of their construction and demolition; of families raised and of people working, farming, and migrating; none of this ever happened. The Negev Bedouin themselves fear that they are disappearing. Because without their places, what do they have—what does being Bedouin amount to? Place isn't a stage, a backdrop against which we act out our lives. It is part of what we are.

Traffic Island

54° 58' 52" N, 1° 36' 25" W

I am staring at a triangle of land surrounded on all sides by steel crash barriers and busy roads. Two corners are covered in bushes and saplings but the center and the sharpest end, which is under an overpass, are stony and bare. This unreachable traffic island is on my walk to work, which for about five minutes takes me alongside a section of inner-city motorway. It's visible through the wire mesh that fences in the motorway, a semi-verdant kingdom that features on no maps. It seems pristine—two wide-screen TVs, ends of carpet, and some odd gunk in a plastic bag have been dumped behind undergrowth along my side of the fencing, but over there, beyond contact, I can see only infant trees and gravel.

These places are everywhere, part of everybody's geographical routine. They are easily ignored, but once you start noticing any particular one it can start to exert a queasy fascination. It's as if you are seeing a landscape that is invisible to everyone else, a secret and intimate kingdom surrounded by unseeing people. This one is in Newcastle in the northeast of England, on a 1.1-mile-long motorway, the A167(M), which opened in 1975. The A167(M) can be a challenging drive, even for those who know the city. Around the triangular island cars nudge from slow on-ramps into dense traffic traveling at up to 70 miles per hour. Some of the merging vehicles then have to cross three lanes of traffic to get to their exit, a mere hundred yards or so farther on. It's a landscape of clamped teeth and grim intensity. There is no time to see anything other than what you might hit or what might hit you.

The triangle is a remnant. The roads were thought about, carefully plotted, and justified, but this island simply happened. This isn't true of all the motorway's green spaces: the roundabouts are just as inaccessible, but they were planned and are dutifully planted and mowed and sometimes sport bulky items of public art. What marks out places like the triangle is the absence of any discernible will either to shape or to create them. They have a quality of abandonment but also of independence, of autonomy from the motorized anthill that is the modern city.

My triangle doesn't feature in *Diversion,* the newspaper produced by the Newcastle City Council in the early 1970s to win over the local population to the idea that their neighborhood was soon to be plowed through with expressways. The editor of *Diversion* appears to have been convinced that the virtues of multilane highways in the inner city spoke for themselves. The harsh line drawings of overpasses that dominate its front page look almost as unappealing in print as they are in reality. The newspaper's attempts to soften the blow were brazenly tokenistic: "1,000 new trees will be planted," along with "22,500 shrubs of varying species," and the dugout earth will be made into a ski slope. That last promise did come true: for a few years what remains a large bump in a nearby park was labeled a ski slope on city maps. It was never used for this purpose, however, since Newcastle, then as now, gets little snow.

Arial photos from the 1960s show that the area that now includes the triangle was once occupied by a school field and long rows of Georgian houses. Both field and houses are still there but bitten off, ending abruptly before the twin-level highway. The violence and suddenness of the transition created a deep sense of loss. The schism between the past and the new world that was built over it has never healed. Almost as soon as the highway was completed, community projects, and now websites, began to be created that gather together pictures, maps, and recollections of the place that was knocked aside.

The unnameability and arbitrary nature of modern remnants like the triangle seem to mock the old streets, but the meaning of these offcuts is inherently amorphous, forever open to reinterpretation. Today in some cities there is a vogue for the most accessible versions of such places to be named, even micro-farmed and semi-inhabited, although not yet in Newcastle, a city immune to such bohemian habits. The postindustrial creative imagination circles these scraps; they suit the academic fascination for transgressive in-betweenness. It is a fashion that has spawned a slew of neologisms among postmodern geographers: "dead zone," "nameless space," "blank space," "liminal space," "urban void," "terrain vague," "gapscape," "drosscape."

But such places are too legion to be co-opted by academic jargon. The only writing that really stays with me when thinking about my traffic triangle is a novel. J. G. Ballard's *Concrete Island* is about a man called Robert Maitland who, after a car crash, finds himself marooned on just such a place: "Maitland saw that he had crashed into a small traffic island, some two hundred yards long and triangular in shape, that lay in the waste ground between three converging motorway routes."

Concrete Island is Ballard's journey into the psychological damage and opportunities of the contemporary landscape. It hardly matters that Maitland's would-be rescuer, Jane Sheppard, has no trouble clambering away. Maitland is stuck because the island induces in him an ever more desperate desire to create meaning out of placelessness. He has to stay in order to create rituals, naming and declaiming over the separate regions of his new domain like "a priest officiating at the eucharist." "I am the island," he declares.

Elsewhere Ballard writes, "Rather than fearing alienation . . . people should embrace it. It may be the doorway to something more interesting." Yet I do fear alienation, and with good reasons. One is my daily journey past this traffic island in Newcastle. It offers a trauma deeper even than the utterly personal one that is charted in Ballard's *Concrete Island* because it is a place emptied of so *many* histories. The sliced-off terraces and fields look as if they have been freshly cut. It's a mutilated landscape, somewhere to look away from, far easier to ignore than acknowledge.

Could I claim this island, become a thirty-minute Crusoe amid the din? Perhaps it's the only way I can get this place out of my mind and stop this possibly unhealthy obsession. And there might be something there, a hidden structure, or hatch, something left from the past. It has become a necessary trip, and I've chosen a relatively quiet late morning to make my pilgrimage, the only daylight period when it is possible to get across the traffic. The on-ramp isn't too bad, and the safety barrier before the island is a bit buckled in one spot, allowing knee-height access. But as soon as I'm over it and onto the island I feel acutely self-conscious. There are a variety of young maples and alders and other self-seeded shrubbery. As the traffic swarms about me I attempt to look purposeful, like a council official surveying biodiversity, doing something meaningful. But I quickly realize that while there may be other, more welcoming traffic islands to explore, this one is beyond human recuperation. There will be no "officiating at the eucharist." In fact, I have a strong desire to lie flat and disappear from view. I'm stopped only by the knowledge that I will then become immediately visible as a dumped body.

After five minutes I'm safely back on the mainland, bubbling with nervous energy. I have come to the conclusion that this particular island isn't nameable or knowable and it cannot be imaginatively reclaimed. Not by me anyway. It retains its dignity, but I have somehow lost mine. There are plenty of other "gapscapes" where I might have better luck. The recolonization of the city still seems like a necessary task, but for the moment I'm keen to get away. I stride purposefully back to safety and in a few moments I'm away from the on-ramp and breathing more easily.

DEAD CITIES

Wittenoom

22° 14' 10" S, 118° 20' 08" E

Places without people are paradoxical. Though they appear functionless, they often have considerable symbolic power. The starkest examples are empty towns constructed for political purposes, but places emptied because of conflict and environmental disaster can be equally potent.

Thirteen hours by car from Perth in Western Australia, Wittenoom is a cursed place. In 2007 the town officially ceased to exist. Wittenoom was a blue-asbestos mining town. According to the state government, Wittenoom is still contaminated with carcinogenic fibers. It has been taken off the map to join a global list of one-industry towns destroyed by their own industry.

Although there is a long tradition of viewing calamity in geographical terms, we have become increasingly nervous about being reminded of failed or "fallen" places. The journey from Sodom and Gomorrah to the disastrous places of the late industrial age is also a shift from the strident, hectoring religious geography of the premodern world to a culture of avoidance and unease. Where once bedeviled cities were constantly invoked reminders of the ceaseless ingenuity of evil, the poisoned places of the secular era are hidden from view. The transition from one kind of moral geography to another deposits us in the deleted town of Wittenoom.

About twenty thousand people lived at Wittenoom before the mine was shut in 1966. The town was taken off the power grid in 2006 and the state government in Perth issued dire warnings to anyone thinking of going there. The official decree maps out a plan of erasure:

- the town of Wittenoom should be closed as soon as possible;
- all buildings and structures in Wittenoom should be demolished and associated infrastructure removed to remove any easily visible sign of past habitation;
- road access to Wittenoom and Wittenoom Gorge to be reviewed with a view to realignment, or closure and removal.

Western Australians expected no less. Contact with any form of asbestos can be fatal, even after the briefest of exposures, and blue asbestos is the most deadly variety. Hundreds of Wittenoom workers, residents, and even casual visitors have died of mesothelioma and other asbestos-related illnesses. The subtitle of *Blue Murder,* investigative journalist Ben Hills's 1989 book on Wittenoom, is *Two Thousand Doomed to Die.* Although much of the town had already been demolished, an entrepreneurial cussedness sustained Wittenoom into the 2000s. It hung on for years, with a population of about thirty, as a ghost town curiosity (the bumper sticker available from the town's souvenir store read, "I've Been to Wittenoom and Lived"). Compared to the dead silence of what was once the world's largest blue-asbestos mine, the totally deserted town of Koegas in South Africa (closed in 1979), Wittenoom was almost lively. In recent years, however, the number of permanent residents has fallen to only five, and the state government is now determined to move everyone out.

I first heard about Wittenoom on a health and safety film featuring the Australian entertainment legend Rolf Harris. He set off for Wittenoom in 1948 pursuing an idea to paint the area's spectacular gorge scenery. Finding that he wouldn't be allowed access to the gorges without signing up to work, he became, in his own words, an "utterly useless" miner. Crawling in the low tunnels, Harris experienced at first hand the mine's minimal safety standards and the "haze of dust" around the rock-crushing area. Luckily for him he found the work impossibly backbreaking and didn't stay long. Instead, it was his father who died of asbestosis. This may have been caused by his work at a Perth power plant or when he built the family's "fibro shack," an Australian asbestos-walled kit house popular during the postwar building boom.

Mining for asbestos in this remote region began in 1938. An upsurge in demand during the war years saw activity expand until, in 1947, the company town of Wittenoom was built to service the mines up the gorge. By the 1950s it was a considerable settlement, but profits were falling since Wittenoom could not compete with the giant South African operations. Its closure in 1966 was more a reflection of the fact that it was running at a loss than dawning health concerns. It was

only by the late 1970s that Wittenoom was fully exposed as
the worst industrial disaster in Australian history.

The Western Australia government considers the cost of
cleaning up Wittenoom to be prohibitive, and it is also understandably nervous about the idea of luring people back to
a place where new hazardous waste sites might yet be discovered. So Wittenoom is deleted. This is a treatment regularly
meted out to disaster towns. They are not merely closed down;
all mention of them is removed from signposts, postal direc-
tories, and official gazetteers. The roll call of poisoned towns
includes such places as Pripyat (see page 114), the now largely
abandoned town that housed the workers of Chernobyl; Bech-
vovinka, a Russian nuclear submarine town, deserted because
of radiation leaks; Centralia, a mining town in Pennsylvania
made uninhabitable by an underground fire that began in 1962
and is still burning today (the road into town bears the graffiti
legend "Welcome to Hell"); and Gilman in Colorado, a lead-
mining town closed because of ground toxicity.

Despite their lowly profiles, these places can at least be
name-checked. They might even be said to be famous, if only
when compared to the many thousands of smaller, less distinct pockets of land that have been contaminated and closed
off. We all know of such sites. I don't have to walk far from
my front door in Newcastle to find large tracts of city land
poisoned by lead, arsenic, cadmium, and zinc. Contaminated
ground is common. Sometimes these places acquire a local
reputation, becoming landscapes of intrigue for the adventurous and the timid alike. They symbolize the evils of un-
checked industrialization for a few, but for most they seem to
evoke something more diffuse, a kind of generalized dread.

For governments having to cope with these stains on the map, vanishing them away is the obvious and easiest solution. Yet while all the erasing and banishing that goes on around us has a solid health and safety logic, there are other human needs to consider. I'm not just talking about the need for messages that remind us of environmental tragedies but something more universal and much older. After all, the desire to morally organize the landscape goes back a long way. Geography was once central to morality and religion. Heaven, Hell, and all the other destinations and journeys of salvation and damnation were understood as permanent places and cartographic realities. They offered a moral map that helped people situate themselves in an ethical landscape. Hell was below, Heaven above. Such literalism may sound quaint to modern sensibilities, but it seems that we still need morality to be tied down and rooted to particular places and specific journeys. If our moral categories float free from the earth, they float away. Religion has always been upfront about all this, meeting the understandable need of earthbound creatures for moral questions to be written into the hills, and for salvation to be a physical destination.

So rather than being deleted from the map, places like Wittenoom should be kept before us as visible manifestations of the consequences of greed and ignorance. They are parts of our lives, of our civilization, and they should be acknowledged with a steady and remorseful determination. Abolishing them leaves us with a deceptively and unconvincingly airbrushed landscape. Wittenoom should be treated as a memorial and paid the kind of attention currently reserved for battle sites, albeit from a safe distance.

Kangbashi

39° 35′ 59″ N, 109° 46′ 52″ E

As we have come to see, places have power, and power is symbolized by its possession of place. The deep bond between the two is especially clear in empty landscapes, such as the ghost town of Kangbashi New Area in the Chinese city of Ordos.

I first heard stories about Kangbashi, a newly built empty quarter set in the arid landscape of the Chinese province of Inner Mongolia, in 2009. Journalists described streets lined with tall apartment blocks, grand plazas stuffed with iconic architecture, and yet not a soul in sight. Since then, other Chinese ghost towns have been spotted, often from satellite images. Huge new towns or suburbs freshly built, apartments waiting for residents, museums without visitors, shopping malls with no shoppers. Inevitably the accompanying news stories were headlined with dire prognostications of housing bubbles and financial meltdown. But this story is stranger than that. On closer inspection it turns out that Kangbashi is a latter-day provincial version of imperial geomancy. A local government has built itself a grand mini-city in its own honor, designed to inscribe and secure its power both through and on the landscape. At the heart of this new zone sits the vast palace of the borough. Tree-lined avenues radiate from it. It bulks out over the surrounding landscape, the throne of a municipality that has become fantastically rich very quickly.

Ordos, a Mongolian word that means "many palaces," is at the center of China's coal country. The city's GDP grew from just under $2.5 billion in 2000 to $41 billion by 2009. It is a frontier boomtown where serious money is being made while the surrounding plains are being tunneled by anybody brave enough to risk the rewards. The landscape across the wider region is freckled with thousands of pits. Many are small-time operations where miners squeeze themselves down unsupported shafts that are not much bigger than they are. Underground fires are common, and many have proved impossible to quench. It's dangerous work but the rewards can be huge. And the tax revenues and payments for licenses pour into Ordos. All that money has created a towering sense of municipal ambition. The city of Ordos wanted to turn itself from a backwater into something magnificent.

Even calling itself a city is a daring claim, since most of Ordos is grassland. On its eastern side sits the old town of Dongsheng, with its narrow dusty lanes. You have to travel 25 kilometers south to get to the ghost zone of Kangbashi but it is all Ordos, a "city" vast in size but with a population density of a mere 18 people per square kilometer — by comparison, London has nearly 5,000 people per square kilometer and Manhattan 25,000 per square kilometer. Ordos doesn't need to play by familiar rules or worry too much about the consequences. The city bosses shifted nearly 400 rural families to clear the site for Kangbashi. It was built for 300,000 people, and it was not built on the cheap. It offers top-end apartments, lavish public squares, parks, and two man-made lakes that stretch for kilometers.

By 2010 city officials had to admit that fewer than 30,000 people had moved in. Conspiracy theorists in the West detect a master plan, and a story went around that, foreseeing global

Armageddon, Communist Party bosses in Beijing had ordered the construction of a ring of cities to house key population groups if the rest of the country suffers nuclear attack or is flooded. One US conspiracy website explains that "for the communist Chinese government, building a network of new cities in strategic locations (such as the Mongolian highlands) to house hundreds of millions of refugees would be a very wise plan." A more plausible explanation for Kangbashi is that local party officials got their planning wrong. They did not anticipate the scale of real estate speculation their project would unleash. All over China the new-moneyed middle classes with spare cash bought up properties in Kangbashi with the intention of making a killing but no intention of moving in.

Yet Kangbashi isn't just about economics. After all, the city kept being built even when it was obvious it was going to stand empty. The scale of its structures, the size of its parks and squares, make no economic sense and never did. This is also a story of a local government that came to see itself as the Yellow Emperor of the coalfields and set out to use the landscape, as emperors have always done, to sustain their authority and announce their permanence.

From their civic palace, surrounded by empty streets and empty museums, Kangbashi's rulers lay claim to a long imperial tradition of civilizing the wastes. It is no accident that the city is planned around a north-south axis, the old imperial urban pattern, nor that the municipal buildings have water to their front and hills to their rear, for in traditional geomancy this is the most auspicious combination. The symbolism extends outside in Sun Square, which stretches two kilometers down to a lake. The square is lined with grand cultural statements such as the shiny aluminum blob that is the Ordos museum as well as an elegant and airy library and an arts center. It is all done on a vast scale. And the fact that the city is populated by only a few disconcerted tourists hardly matters. These monuments to creativity and learning express and confirm the place of power just as surely as sacred paths and temples once did in ancient China. Such is the faith invested in this site that it is even imagined to resolve, or at least displace, political conflicts. In Kangbashi's Genghis Khan Square sits a huge statue of the Mongol warlord and other plinthed stone giants that nod to Mongol identity, such as the two rearing horses energetically clashing hooves. The often tense relations between Mongols and Han Chinese, who comprise 90 percent of Ordos's population but are far less dominant across the rest of Inner Mongolia, are resolved into shared symbols of ambition and entrepreneurial drive. Any trace of the filthy work that generates the money needed to pay for all this elaborate architecture is nowhere to be seen.

Building peopleless cities has become something of a habit among Chinese urban planners. There is the empty desert town of Erenhot, Zhengzhou New District, and many other resident-free settlements that have not yet been given a name. The Chinese have also built one in Angola. Kilamba New City, thirty kilometers from the capital, Luanda, is designed for half a million people. It has a dozen schools and more than seven hundred eight-story apartment buildings but it stands empty.

For all their confidence, these conjured landscapes have an urgent, almost desperate quality. Empty cities can evoke power but they cannot secure it. They point to the vulnerability of authority even as they act out its overweening will.

Despite the planet's rapidly growing population, the early

twenty-first century may well be known by future urban geographers as the era of empty cities. At no point in the past have we seen the construction of so many and on so magnificent a scale. Any media coverage they get is bewildered and, in the case of China, full of schadenfreude, one of the few pleasures left to an envious world watching and wondering at that country's spectacular urban growth. To understand Kangbashi it is necessary to know that it is as much a symbolic landscape as a practical one: it is a new urban form that draws on old magic.

Kijong-dong

37° 56′ 12″ N, 126° 39′ 21″ E

Kijong-dong is a fake place where the lights go on and off inside tower blocks that have no glass in their windows. There are no residents and no visitors are allowed. But the lights are on timers and the roads are periodically swept clean. Kijong-dong, which is also called Peace Village, in North Korea, was built in the 1950s to lure potential defectors from the South and as a display of the communist state's progress and modernity. The question is, what remorseless logic keeps it going?

Full-scale simulated cities are rare. They are sometimes called Potemkin villages, after the Russian minister who supposedly had fake villages built, complete with glowing fireplaces, in the recently conquered lands of the Crimea. It is said that he hoped to convince Catherine II that this was a prosperous and well-populated land. Unfortunately, there seems to be little truth in this legend. Better examples come from the Second World War, when decoy towns were quite common. One of the largest was a fake Paris, built to attract enemy bombers away from the real city (see "Arne," page 12). But this was a hasty job, gimcrack in comparison to Kijong-dong. The idea of a permanent fake civilian village, deployed to make people across the border think things are going well, seems to be uniquely North Korean.

Peace Village is a product of the armistice treaty signed in 1953 between North and South Korea. A 4-kilometer-wide demilitarized buffer was established between the two nations and each was permitted one settlement within this 250-kilometer-long no man's land. The South decided to retain the rice-farming village of Daeseong-dong. The North Koreans chose to build Kijong-dong directly opposite it, about a mile across the frontier. It was a much larger place, and images from Google Earth show a sprawling town comprising three main centers, interspersed with farmland. Each of the centers has rows of what appear to be very large houses or public buildings, many with large gardens. Although it does not feature on many maps of the country, Kijong-dong was built to impress. The costly blue-tiled roofs on many of the concrete buildings and the electric power supply proclaim an anachronistic vision of luxury and success. In the context of the thatch-roofed peasant buildings typical of the area in the 1950s, Kijong-dong must have looked like the future. At the time, mass housing and electrification were symbols of communist progress, but it is unlikely that observers from south of the border find them impressive today. They know that

North Korea is poor and that it is one of the least illuminated countries in Asia. Nighttime satellite photographs show it as a pitchy emptiness surrounded by brightly lit neighbors.

The official North Korean position is that Kijong-dong is a thriving community; that it contains a large collective farm (run by two hundred families) and many social services, such as schools and a hospital. Yet Kijong-dong is so close to the border that, with the aid of binoculars, people can see it is empty. And plenty of people do. During lulls in the level of hostility between the two countries, the border crossing draws a steady flow of tourists. They are eager to step across the demilitarized zone into the rarely visited nation to the north. Visitors, who are warned not to make eye contact with North Korean soldiers or gesture at them in any way, are taken to the nearby village of Panmunjom, from which Kijong-dong is even closer, clearly visible in the distance, though it is still very much off-limits. Panmunjom's only attraction is the pleasure of straying into a forbidden zone. Tourists may also thrill to the official South Korean warning that their little journey across the border "will entail entry into a hostile area and possibility of injury or death as a direct result of enemy action."

Other, newer propaganda tools also compete for the skyline. A nearby 525-foot North Korean flagpole, erected in retaliation for South Korea's putting up a 323-foot flagpole in Daeseong-dong, was, for a while, the world's tallest. Yet Kijong-dong remains a potent and, until recently, noisy symbol. Until 2004 loudspeakers on its empty buildings pumped out denunciatory speeches and patriotic operas across the fields almost every hour of the day and night. After a few years of silence, in 2010 the speakers went back on, not long after the North Koreans had sunk a South Korean submarine, killing forty-six of its crew.

Kijong-dong may seem like a novelty, but it is part of a twentieth-century tradition of hollow architectural spectacles. Communist regimes from Moscow to Beijing often indulged in monumental and monumentally useless buildings. They were built as expressions of revolutionary zeal and the permanence of the new order. What are we to make of the 1,100 rooms of Bucharest's Palace of Parliament (a.k.a. the House of Ceaușescu), the second-largest building in the world, which was still being furnished when Nicolae Ceaușescu was thrown from power in 1989? Or Bulgaria's Buzludzha Monument, a vast spaceship-shaped tribute to communism, filled with garish murals, that sits, remote and inaccessible, on the top of a mountain? Kijong-dong is part of a long tradition of clumsy architectural propaganda. It is a tradition that celebrates symbolism over utility, gesture over substance. It seems desperate for everyone to admire it but only at a distance — it's a psychopolitical complex that doesn't just spawn fakes but lovingly maintains them.

Across North Korea, monuments to prosperity and progress abound. The country is home to an Arc of Triumph, the largest arch in the world, which stands over a mostly empty highway. Built in 1982, the arch is inscribed with the "Song of General Kim Il Sung" and made up of 25,550 bricks, one for each day of Kim's life. There are also the vast stone women that make up the span of the Three Charters for National Reunification monument, which yawns over another empty road. High above the capital, the 170-meter Juche Tower com-

memorates the seventieth birthday of the man who brought
the country to its present parlous state, Kim Il Sung. It looks
down at military parades during which fake missiles are trun-
dled out for the benefit of an admiring world.

As part of their unsuccessful efforts to cohost the 1988
Olympics, held in South Korea, the North Koreans also built
cavernous and little-used sports arenas. In the capital, Pyong-
yang, Chongchun Street is lined with a huge table tennis stadium, a handball gymnasium, and a tae kwon do hall. Most
spectacular of all is the 105-story Ryugyong Hotel, one of the
world's largest hotels and the tallest building in North Korea. Its colossal pyramid shape dominates the capital. Building started in 1987, but it is still not finished, and it is unlikely
that the hotel will ever attract the foreign tourists or investors
it was supposedly designed for. It is another fake, a nostalgic
ruin of the future that pretends, like Kijong-dong, to want to
lure us in but actually doesn't want anyone anywhere near.

Ağdam

39° 59' 35" N, 46° 55' 50" E

Ağdam is the world's largest dead city. It is a place of ruins.
Around the central mosque, one of the few buildings still with
a roof, stretches a scene of destruction. If you peer down on
Ağdam from Google Earth, you would be forgiven for thinking that a nuclear bomb had just exploded.

Ağdam was a casualty of a war over the nearby ethnic enclave of Nagorno-Karabakh (Karabakh for short), between Azerbaijan and Armenia and their ethnic allies, and was deserted then systematically blown apart in 1993. Between 1992 and 1994 many thousands died across this region in brutal ethnic battles while the rest of the world was still celebrating the fall of the Soviet Union. Karabakh saw some of the worst bloodshed. From the start, war reports were being filed of the scalping, beheading, and mutilation of civilians, including children. Along with the dead, three million were displaced, among them all of those who once called Ağdam home.

The attempt to destroy places entirely is a characteristic of modern, total warfare, in which the destruction of the enemy's will to fight is assumed to follow from the annihilation of its centers of civilization and civilian life. The level of bombardment is a perverse tribute to the primary role of place in human identity. Typically, only the center of town is destroyed, and is quickly rebuilt after the war's end. The complete nature and the longevity of Ağdam's ruination make it distinctive, as well as the fact that it was so recent. A few decades ago Ağdam was a busy regional capital, known for its lively bazaar and for its quaint bread museum. Although a predominantly Muslim city, it was also renowned for its wineries. The Ağdam Brandy Company was rooted in a century-long tradition. Today one of the city's many ghosts is a brand of fortified wine, still drunk in some of the former Soviet states, known as Ağdam. It's the kind of potent and cheap brew that Russians call a "mumble juice."

Ağdam doesn't get many visitors. The accounts of the few travelers who have made it past the minefields describe an apocalyptic landscape. Here are brief extracts of two such tales from travel bloggers, the first from Justin Ames, the second from Paul Bradbury:

> One thing that captures your attention before long is the scale of the destruction. Every time you think you are near the edge of the city or the end of a road, you go over another hill or around a bend and a whole new field of destruction opens up in front of you.
>
> In this former town of 50,000 people we saw fifteen civilians (a mother and two sons were picking berries, which were growing wild in the main street; an elderly couple with granddaughter were foraging for firewood; the others were collecting scrap metal) . . . On one broken gate I saw the number 50. House number fifty, but which street? No other identification was evident. Even the roads had been dug up and all the pipes removed.

What is most telling about these accounts is their tone of surprise. "I had never even heard of Ağdam," they chorus. How many of us could say otherwise? Or where in the world Karabakh might be? As the ragged fringes of the former USSR have been transformed into a shifting delta of enmities, the outside world has come down with shock-induced geoamnesia. In North America or Western Europe the region's place names, unpronounceable, unplaceable, fly up every so often out of the news but are instantly forgotten. For anyone over a certain age it is hard to believe that we utterly mistook something so big, so solid, as the USSR. Even at a distance of almost a quarter of a century it is difficult to grasp that it was never a country at all but an unwieldy empire.

Ağdam is an endless source of surprise, not least because it keeps on getting forgotten. For the Armenians and Azeris, the ethnic majority in Azerbaijan, by contrast, it is an ever-potent reminder of what both sides like to cast as an age-old territorial dispute between victim and aggressor. The fiercely Christian Armenians have good reasons to think they are surrounded by enemies. Azerbaijan, to their east, is a Turkic state and stalwart ally of Turkey, its western neighbor and a country that not only denies responsibility for the Armenian genocide of 1915 but prosecutes people who talk about it in public, under a law that forbids "insults" against the Turkish nation. But Turkic peoples across the region have their own history of persecution and genocide. They have been the victims of countless ethnic massacres, including in Karabakh.

When the Bolsheviks finally took control of this warring region in the early 1920s, they began to make deals to win over the biggest national groups. They first promised Karabakh to the Armenians, but then, in order to win over Turkey, gave it to Azerbaijan. Simmering disputes were not so much resolved as crushed. These conflicts became increasingly public in the late 1980s when protests throughout Karabakh called for an end to Azerbaijani control. The Kremlin refused to countenance any change in the status quo, but when the USSR fell apart, so did the brakes on ethnic conflict.

Ağdam was singled out for special treatment in these disputes because of its strategic location near Karabakh. But it was also targeted because it provided the backdrop for street protests against the breakaway of Karabakh in the late 1980s. In 1988 street fighting erupted between ethnic Azeris and Armenians in and around Ağdam. The town became a symbol of Azeri militancy and resistance. It was the memory of this that seems to have spurred the pro-Armenian Karabakh army to such vindictive destruction a few years later. Certainly, the Karabakh army's explanation was weak; it claimed that Ağdam was being used as a military base. Yet the city was poorly defended and soon fell, its population fleeing before the invading troops. The invaders then withdrew and subjected the empty city to continuous artillery bombardment until almost every building was destroyed. Other regional towns were also attacked, but the assault on Ağdam, because of its size and thoroughness, remains remarkable. Today the war is in abeyance, but there are few signs that the conflict is over.

The Armenians know very well that most atrocities get forgotten. Karabakh declared its independence after its military success, and although it is not recognized by any other country, even by its kin nation of Armenia, it is a de facto sovereign state. The Karabakhians' position is that they will hold on to Ağdam, and the rest of what they call their "security belt," until their independence is recognized by Azerbaijan. But recognition is a distant prospect, and in the meantime, all that the Azeris have to hold on to is a different label not only for the "security belt" but for the whole of Karabakh: it's "Armenian-occupied Azerbaijan."

The legal claims are far apart while the ruined city crumbles away. In the interregnum before the next wave of violence, small steps to rehabilitation have been made, and in 2010 the Karabakh government announced that the central mosque had been partially restored. However, the propaganda mileage that can be gained from restoring one building in the middle of a wrecked city is small. And even this gesture has elicited bitter recrimination. The media in Kara-bakh has reported vox-pop reactions such as, "If the Azeris destroyed our cemeteries and churches, why are we restoring their mosques?"

Both the total nature of the destruction of Ağdam and the long time it has remained deserted have established it as an iconic site of suffering and anger. The grief is compounded by the inability or unwillingness of the outside world to no-tice Ağdam's, or Karabakh's, calamity. The only real rebuild-ing of Ağdam that has taken place has been symbolic, among supporters of the Ağdam soccer team. Imaret Stadium, built in 1952, was home to FK Qarabagh Ağdam, but the building was destroyed along with the rest of the city and the team scattered. The team has since become a cultural symbol to the region's Azeri refugees, and with financial aid from both Turks and other Azeris the club was reborn and now plays in the Azerbaijan Premier League. It is one of the most suc-cessful clubs in the country and has made it through to Eu-ropean-level competition several times. A new "home ground" in Azerbaijan has been found for it. Its success contrasts with the fortunes of what was once one of the USSR's top clubs, FK Karabakh Stepanakert, which is based in the Karabakh capital. Banned from the international game, that team has withered, and today it has no money and only local fans to rely on. It sounds like a parable: the ghost town cheered to the rooftops and the decline of the crowing victor. But suc-cess at soccer is thin consolation for the loss of a city. The an-nihilation of place has consequences for both its victims and its perpetrators, and until it is rebuilt the dead city of Ağdam will continue to excite further hatred and violence.

Pripyat

51° 24' 20" N, 30° 03' 25" E

The flip side of urbanization is the fantasy that one day nature will return and the hostile concrete of the city will be carpeted with flowers. But as our capacity to poison the earth has grown, so this dreamscape has turned sick. While it's true that nature has returned to reclaim the Ukrainian city of Pripyat, this is mostly explained by the fact that radiation levels there are so high that all the humans have had to be evacuated.

Less than three kilometers away from Pripyat is the Chernobyl nuclear power plant. Late one day in April 1986 the residents of Pripyat heard the following announcement from their local radio station: "An accident has occurred at the Chernobyl Nuclear Power Plant. One of the atomic reactors has been damaged. Aid will be given to those affected and a committee of government inquiry has been set up." Later the same day the entire population of forty-five thousand was bundled into over a thousand buses, without any time to pack. Clothes were left in wardrobes, toys remained in empty prams, pets were abandoned. People were told they would be away for just three days, but they never came back. Even the local army unit's tanks and helicopters were left where they stood. As we now know, they should have, and could have, gotten out sooner. Reactor No. 4 had exploded three days earlier but the accident was kept secret, leaving the city subject to lethal levels of radioactivity that would have devastating consequences for many of those who lived there, as well as for subsequent generations.

After 1992 and the collapse of the Soviet Union, on-site security lapsed and Pripyat became prey to looters, who even stripped out all the wiring and linoleum. But while human life abandoned the city, over the coming years nature surged back. Today the roads and buildings have been cracked open by the roots of young trees. Mosses and grasses cover the asphalt and decaying concrete, and as the city's drainage system clogged, after each spring thaw paved areas become shallow lakes. An amusement park, complete with Ferris wheel, due to open on May 1, 1986, stills stands, turning to rust amid the weeds.

An old dream has come back to mock us. In 1890 William Morris wrote *News from Nowhere,* in which he delighted in a vision of the city drawn back into nature. He prophesied Londoners turning against the ugly streets and creating "a very jolly place, now that the trees have had time to grow again since the great clearing of houses in 1955." It was a powerful idea and tapped into a growing sense of urban malaise. As the world has been covered in increasingly large urban conglomerations, the desire to see nature take its revenge has become ever more thrilling and dangerous. But Morris, the anti-industrial prophet, could not have imagined that nature's revenge would look like this. It is estimated that Pripyat will be safe again for human habitation in about nine hundred years. Radiation levels are so high that even the briefest of visits is ill advised, and the exclusion zone around the site, officially known as the Zone of Alienation, covers 2,600 kilometers, which is bigger than Luxembourg. The most dangerous places are inside the buildings, where contaminated dust and debris have settled. Jill Dougherty, an American journalist based in Moscow, recalls a drive around Pripyat: "It is completely quiet — it is the most eerie experience I have encountered." She goes on to describe pavements that "have been taken over by moss and brushwood" and "houses literally rotting . . . I could hear the sound of dripping water coming through the ceilings."

When it was first built, Pripyat was a model Soviet town. Building began on February 4, 1970, and "shock construction" rapidly created a home for numerous Soviet nationalities. The street names — Enthusiasts, Friendship of the Peoples — reflected Pripyat's diversity. It was a bright city of wide streets and modern apartment blocks, many decorated with ceramic tiles. The average age of the residents was only twenty-six, and more than one thousand babies were born every year. One former local recalls with pride, "Only in this city could you see a parade of children's strollers, when in the evening, mothers and fathers walked the streets with their babies."

For a while it looked as if nothing would survive the world's worst nuclear accident. In the immediate aftermath of the blast everything was affected, often in odd and gruesome ways. Animal embryos dissolved, and the thyroid glands of horses literally fell apart. One large area of pine woods in the path of the fallout became known as the Red Forest as the trees changed color and died. Today, though, the forest is green again. Many plants adapted rapidly to the new environment. A comparative study of two plantings of soybeans, one sowed five kilometers from the reactor and one sowed one hundred kilometers away, has found that the former were highly contaminated and weighed half as much as they should but also that they were undergoing molecular adaptation. For example, they had three times as much of an enzyme (called cysteine synthase) that protects plants against environmental stress as normal plants.

Meanwhile, the city and its surrounding exclusion zone have been colonized by a variety of animals. Radioecologist Sergey Gaschak has observed that "a lot of birds are nesting inside the sarcophagus," the concrete shell that was built over the blown reactor in 1986. At the epicenter of the disaster he has spotted "starlings, pigeons, swallows, redstart—I saw nests, and I found eggs." A head and species count of fauna in the exclusion zone in the mid-2000s found 280 species of birds as well as 66 species of mammals, with a total of 7,000 wild boar, 600 wolves, 3,000 deer, 1,500 beavers, 1,200 foxes, 15 lynx, and thousands of elk. Bear footprints have also been spotted. This is something of a revelation in this part of Ukraine, since bears have been unknown here for many years.

Mary Mycio, whose *Wormwood Forest* is a widely read natural history of the site, argues, "On the surface radiation is very good for wildlife." The reason is simple: "it forces people to leave the contaminated area." Referring to the wider exclusion zone, she claims, "It is a radioactive wilderness and it is thriving."

Yet it would be odd if Pripyat and its environs were harmful only to human beings. Another way of looking at the area is as a zone of mutant nature. The flora and fauna may look like they are "thriving," but that's only by means of a crude head count in comparison with normal cities. Timothy Mousseau, a professor of biology at the University of South Carolina who has studied the area in depth, conceded to *National Geographic News*, "One of the great ironies of this particular tragedy is that many animals are doing considerably bet-

ter than when the humans were there." But he also warned
that "it would be a mistake" to conclude that this means they
aren't suffering. In fact, Mousseau's research shows that re-
productive rates among local birds are much lower than average, and evidence from other studies reveals hormone damage in trees, many of which have been growing in strange
and twisted ways. The mutation of the trees' growth recep-
tors means that, as a colleague of Mousseau, James Morris,
explains, they "are having a terrible time knowing which way
is up." Other work has shown even odder reactions, such as
freshwater lake worms switching from asexual to sexual re-
production.

It is hard to know if these changes are signs of damage or
adaptations or both, but they tell us that this is no Garden of
Eden. When he visited the area in 2005 the Ukrainian president, Viktor Yushchenko, floated the idea that it could become a nature reserve. Since then the local authorities have
been scoping the idea of a Chernobyl National Nature Park.
Paradoxically, at the same time the president proposed that
the site could be used to store foreign nuclear waste. This idea
was soon thrown out, but it drops a heavy clue that Ukraine is
trying to find an economic use for the exclusion zone. All the
"good news" about abundant flora and fauna is being used to
suggest that the area is bouncing back and that lethal radiation and biodiversity are happy bedfellows.

The dream of the city returned to nature is a persistent
one. The more we urbanize and rid ourselves of nature, the
more it haunts us and the more likely we are to take a strange
delight in seeing sidewalks and buildings broken apart by tree
roots. This is what is happening at Pripyat, but the dream
wasn't meant to be like this. William Morris's hope was for

a balanced relationship between people and nature. In 1890 that could still have happened, and one day, perhaps, it might yet be possible. In the meantime, the overgrown streets of Pripyat symbolize the abandonment of that hope. We were supposed to be part of this story: regaining something, rejoining something we'd lost. Pripyat points to an alternative future.

The Archaeological Park of Sicilian Incompletion

37° 43' 37" N, 15° 11' 02" E

Modern places are made up of layers of incomplete visions of the future, and the result is a permanent state of impermanence. Giarre, a small Sicilian seaside town that lies in the shadow of Mount Etna, offers one of the world's most startling concentrations of half-finished grand building projects. This town within a town was dubbed the Archaeological Park of Sicilian Incompletion by Italian artists, and the name has stuck. Here you will find twenty-five incomplete structures built between the mid-1950s and the 2000s, many of considerable size, such as a vast Athletics and Polo Stadium, an unfinished near-Olympic-size Regional Swimming Pool, and a tumbling concrete palace known as the Multifunctional Hall. Their concrete shells are slowly being taken over by meadow grass and cacti, but they still dominate the landscape.

In a town of only twenty-seven thousand people these edifices stand out starkly, as unmissable clues to local politicians'

habit of making impressive but ill-advised claims about what
public works they could see to completion in order to secure
funds from the regional government. Starting large-scale construction work has been a vote winner and a way of creating
jobs. It was also claimed to combat the recruiting power of
the Mafia.

The landscape that has resulted from all these promises
is surreal and has a melancholic appeal for those attracted
to the idea that decay and inertia will always overtake the
hubris of modernity. A collective of artists based in Milan,
New York, and Berlin, called Alterazioni Video, devised the
idea of the Archaeological Park of Sicilian Incompletion in
Giarre, delighting in what, in a photo essay on Giarre, they
call its "sheer scale, territorial extent and architectural oddness." They define incompletion as the "partial execution of a
project followed by continual modifications that generate new
spurts of activity," a process that produces "purposeless sites"
that "dominate the landscape like triumphal arches." Alterazioni Video collaborator and local community activist Clau-
dia D'Aita, who once staged a mock polo match at the Athletics and Polo Stadium, explained to a BBC journalist that
all of Giarre's unfinished edifices should be seen as "a kind
of open-air museum." It's a refrain picked up by Alterazioni
Video, which announced in its photo essay that these "glaring
blemishes on the civic horizon" should be "transformed into a
tourist destination, giving new value and meaning to the monuments of a perpetual present."

Alterazioni Video produced a map and guidebook to help
visitors find their way around the various key sites of incompletion. I'd not heard of anyone using the guide in earnest,
so I went to Giarre in July 2013 to see what it would be like

to be a tourist of unfinished Sicily. It was, unsurprisingly, an odd experience, and I occasionally found it difficult to tell the complete and incomplete town apart. Just across the road from Chico Mendes Park, a half-built and fenced-off "children's city" that is a central stopping point on Alterazioni Video's self-guided tour, is another abandoned area, an elaborate 1980s roundabout that is now a wasteland of grasses, graffiti, and wild fig trees as well as a huge stash of brown glass bottles. It is adorned with a broken central fountain, a rusting orb shaped like *Sputnik,* a ring of dried-up smaller water features, and a weed-infested sculpture of the nineteenth-century cleric Don Bosco instructing street children. As I stood next to Chico Mendes Park, this large traffic island felt unfinished, but it is more likely to have simply not been maintained. Neglect and incompletion merge in Giarre, creating an extensive and continuous landscape of abandonment.

In its "Sicilian Incompletion Manifesto" Alterazioni Video argues that Giarre is the "epicenter" of a phenomenon that has "radiated out from Sicily to the rest of the peninsula, creating an Unfinished Italy." Yet the way the incomplete parts of town mesh with the ordinary landscape reminded me that I didn't need to come to Italy to find the remnants of once heroic architectural visions. Standing in the shadow of the high concrete terraces and walkways of the Athletics and Polo Stadium, on a playing field covered in the ash and cinder thrown up by Mount Etna, I was reminded of my hometown of Newcastle, which has its own network of unfinished concrete walkways and a stub end of a motorway, both discards from 1960s plans to bulldoze the city and rebuild it as the "Brasilia of the North."

Giarre offers the extreme form of a condition found in

most cities, making it a parable of urban planning. It is the epicenter not of merely an Italian but a global phenomenon of accreted unfinished visions. It is also a good place to think about how we live with the layering and churning of the city. Being surrounded by the sawed-off ends of the utopian plans of once powerful people can be liberating, as it subverts the professional's claim on the city; the architects, politicians, and planners all stand defeated, incapable of molding place to their will. Yet if this is a victory, it is a hollow one, for we are all left picking our way through the pieces. A more profound consequence is that we disconnect ourselves from place: provisional and incomplete hometowns inspire provisional and incomplete loyalty. In tumbling together half-realized projects at an ever greater speed, the city of incompletion disrupts the possibility of people building up a relationship of care, knowledge, and trust with the place they live in.

The artists who guided me around the Archaeological Park of Sicilian Incompletion are attempting to find a new and challenging way to reconnect people with place by embracing this sense of disconnect and tumult. It is a paradoxical project, both subversive and conservative, mocking the failure of effective governance in Sicily while suggesting that vaguely futuristic ruins can be the basis for a novel type of geographical allegiance. "The sum of these relics of never-attained futures," they write, "is so vast that it can be considered as a true architectural and visual style, representing Italy and the age in which they were produced." Incompletion comes to represent "the speculative munificence of Sicilians and all other Italians" and, even more grandly, the invention of authentically modern "places for spiritual habitation and contemplation" that are also "places of existential awareness, embodiments of the human soul."

The idea of rebranding the modern ruins of Giarre as the Archaeological Park of Sicilian Incompletion is an attempt to reclaim the contemporary landscape, to allow us to find within its spectacular bleakness both beauty and drama. While the aesthetic of ruins that this argument relies upon looks beguiling as a set of black-and-white photos, on the ground it soon gets wearisome. After I visited a few of the chosen remnants they all started to look the same and I gave up. I'd learned that being a tourist of incompletion has diminishing returns, but I'd also been reminded that cities of incompletion are places that I have spent a lot of time either traveling through or returning back home to.

Dead places are places apart: they have the power to disrupt the way we think about place. Even when tightly controlled, the absence of people means there is something wild, something ungoverned, about them. From dead cities it is but a small step to those places that have somehow escaped the normal uses of territory and seem to be living by a new and original set of rules.

SPACES OF
EXCEPTION

Camp Zeist

52° 06′ 35″ N, 5° 17′ 47″ E

Spaces of exception are places where normal rules do not apply. Often they are created to allow states or communities to follow their own course without interference and beyond the ken of the outside world. Sometimes such places escape the rule of government, but they have also been used by governments who wish to undertake activities that would normally be forbidden, such as "extraordinary rendition" (see "Bright Light," page 136). However, spaces of exception are not only about the breakdown of ethical conduct; they have provided unique opportunities for entrepreneurs and social ideologues and experimenters who want to create a more perfect community. Spaces of exception challenge our notions of sovereignty and ownership, but they also force us to realize how flexible these two ideas can be, a point that is driven home by the story of Camp Zeist.

My interest in the Dutch military facility known as Camp Zeist started in 1999 when it came under Scottish law. It remained Scottish until 2002, at which point it became Dutch once again.

Camp Zeist changed legal nationality in 1999, from Dutch

to Scottish, in order to allow a trial to take place both in Scotland and outside Scotland. The trial was for the two Libyan men who were suspected of bombing Pan Am flight 103, which blew up over Lockerbie in Scotland on December 22, 1988. After fragments of evidence had been pieced together from the crash site, on November 13, 1991, a Scottish sheriff issued arrest warrants for Abdelbaset al Megrahi and Al Amin Khalifa Fhimah. A UN Security Council resolution demanded that the accused stand trial before "the appropriate United Kingdom or United States court."

The Libyan leadership doubted that the men would get a fair trial in either country, so early hopes for speedy extradition were soon quashed. Over the next decade a compromise was arrived at: Libya would agree to extradite the two men on the condition that the trial take place on neutral territory. The Libyans first suggested The Hague but eventually agreed to allow the trial to proceed in the more secure and purpose-built facilities that could be offered at Camp Zeist.

The idea of giving a specific plot of land to another country for a specific purpose might not appear novel. Most of the well-known examples, however, are not quite what they seem. Visitors to war graves dedicated to the foreign dead in Belgium and France sometimes get the impression that they are actually entering Britain, the United States, or Canada, for example. But that isn't the case. Such graveyards are owned and looked after by these nations, but they do not have sovereignty over them. The same is true of the plot under the John F. Kennedy memorial at Runnymede, England, which was "given" to the United States in 1964. Owning a place, or having control over it, is different from making it an integral part of a nation, although drawing a line between the two can be a real headache. Foreign embassies are a classic conundrum. The fact that host states are obliged not to violate embassies' grounds and that diplomats are immune from local prosecution makes them ambiguous places. They remain the sovereign territory of the country they are in, yet they are granted such a raft of concessions that they can resemble enclaves of foreign states. A parallel set of concessions covers many foreign military bases. Guantánamo Bay in Cuba remains part of the Cuban nation, but the United States has a perpetual lease and full jurisdiction over the territory.

Giving up territory to another nation isn't something countries do lightly, and when it happens it is either forced on them or takes place for a defined period of time. Most of the examples that fall into the latter category concern the demarcation not of national but of international zones, and the function of these is often very specific. For example, when Princess Margriet of the Netherlands was about to be born at the Ottawa Civic Hospital in 1943, the Canadian government decided to make the maternity unit international territory, thus allowing her to claim Dutch citizenship through her parents. The "green zones" that harbor and protect military and diplomatic missions in war areas, such as Baghdad's ten-square-kilometer international zone, provide a better-known and more sobering instance.

None of these examples is quite like Camp Zeist, a portion of the Netherlands that for three years became a legal enclave where Scottish law applied. The formal agreement between Britain and the Dutch that set up the trial, signed in September 1998, goes into great detail about the exact range of the court's jurisdiction. It makes it clear that Camp Zeist is being "hosted" by the Netherlands but also that the premises and will of the Scottish court are inviolable. It even sets down who is expected to dispose of the site's garbage and how that is to be paid for and lists all the taxes the Scottish court is exempt from, including "excise duty included in the price of alcoholic beverages, tobacco products and hydrocarbons."

One thousand Scottish police officers were brought in to protect the L-shaped site, which is a little over half a mile long. A Scottish courtroom and prison had to be constructed, as well as a press center, and the court also hosted a vast array of legal advisers. One of the key players was the legal adviser to the US Department of State, David Andrews, who is candid about the fact that the plan to use the Netherlands was worked out between the United States and Britain many months before it was mentioned to the Dutch. "We hoped the Dutch would agree," he recalls in an article for an academic law journal, "but due to concerns over leaks, we felt we could not approach them until we had the entire program between the US and the UK worked out." As it turned out, what Andrews had feared would be the "trickiest and certainly the most crucial part of our initiative: getting the Dutch to agree" was a pushover. As long as everyone was clear that this was a onetime event, the Dutch were happy to play their part in resolving an intractable international crisis. Andrews notes dryly that he had to spend "considerable time" not with Dutch politicians but convincing the band of the Royal Dutch Air Force to temporarily abandon an important part of the site.

The most difficult challenge was the legal and logistical ramifications of creating Scottish territory abroad. The Scottish judges had to be given power not only over the trial but over the court and camp premises. Moreover, Andrews realized that "it would not be practical to absent a group of Scottish citizens from Scotland for the better part of a year." So this was to be a unique type of trial, one without a jury but with a panel of judges. This unusual arrangement required new British legislation, which was rushed through Parliament as an Order in Council, a legislative loophole that allows changes in the law without a vote in the House of Commons.

Another, more local challenge was the road that ran from a nearby aircraft museum to its repair shop, which would cross what was to become Scottish territory. The solution was to build a pair of gates that could be opened to allow museum workers through while cutting off the road to Camp Zeist personnel, and vice versa. The gates worked well, although occasionally people found themselves trapped between them. Richard Bailey, a Scottish court spokesman, was one. "I got caught inside, between the gates once," he told a visiting journalist. "And I did think, 'Where am I?' But mostly I thought, 'When the hell's the gate going to open?'"

On April 5, 1999, the two accused Libyans were flown to the Netherlands and driven straight to a Dutch extradition hearing. After being formally extradited, they were taken to Camp Zeist, where they were arrested by Scottish police. The charge of murdering 270 people was read to them, and they were remanded in custody within the camp. Eventually, after one appeal, Megrahi was convicted and sentenced to life in a Scottish jail in 2002, until he was released on compassionate grounds in 2007. The other man was cleared. Camp Zeist itself returned to the Netherlands and was converted into an immigration detention center.

The UK and US governments regard Camp Zeist as a great success. It put the suspects in the dock and resolved a political headache for a number of states, and it is held up

by some enthusiasts as a model for bringing difficult international cases to trial. Turning a piece of one country into
the legal territory of another for a brief period of time has
been shown not only to be feasible but to speed up justice.
But many who did so much to set up this little bit of Scotland in the Netherlands are more doubtful that this kind of
experiment could or should happen again. After considering
the huge costs and endless legal complexities involved, David
Andrews concludes that "the third country trial is not a model
that we ought to consider lightly, if ever."

Geneva Freeport

46° 11' 18" N, 6° 07' 38" E

Hidden from view, in the dark, there is something growing:
storage places, air-conditioned vaults filling up with an ex-
panding volume of valuable things. This is the antithesis of
our throwaway society. For, at the same moment that consumerism coughs up great rivers of shoddy gizmos to be mashed
up and drained away, it also brings forth increasing quanti-
ties of beautiful and rare objects: paintings, cars, wine bottles, sculptures; items that have to be dusted, photographed,
and catalogued, to be prized and kept forever. Where once
such valuables could be crammed into the houses of the elite,
today their possessions are so plentiful that this is no longer
an option. The relationship between the rich and their objects of infatuation has also changed. Now they buy them as

investments, as an essential component of any serious wealth portfolio.

The Geneva Freeport is a massive, high-spec warehouse of treasure. From the outside it is a nondescript white concrete block surrounded by gray roads and gray parking lots, but it may be the most valuable building on the planet. The total worth of just the art in the freeport has been estimated at $100 billion. Alongside the artworks there are vaults and floors of other rarities, such as three million bottles of wine, decks of expensive cars, even a chamber full of cigars. The vaults are populated by a small army of conservators and inventory takers, but it's a lonely job, often requiring that these experts be shut up inside a safe room for most of the day. One art specialist, Simon Studer, recalls being locked in one chamber that contained thousands of drawings, paintings, and sculptures by Pablo Picasso: "I was checking sizes, condition, looking for a signature and making sure the art was properly measured," he told the *New York Times*. He eventually worked out that the knocking sound from next door came from another inmate who was counting gold bars: "You have no idea what is next door and then you happen to be there when they open a door and, poof, you see."

Freeports are places where goods can be imported and exported free of customs duties or other taxes. They are a medieval invention that have long been useful in easing the flow of trade and were never meant to be sites for hoarding valuables. In 1888 the Grand Council of Geneva voted to establish the freeport. At the time it tended to stock prosaic items and hold them for short periods. But Switzerland's unique regard for the privacy of foreigners wishing to use its tax-free facilities soon began to attract a niche clientele. By the end of the twentieth century the Geneva Freeport had established itself as the world's central repository for a new kind of global investment system based on the buying and selling of objects of high value. Since works can be sold and bought within the freeport without transaction taxes of any sort, it effectively operates as a trading hall. Old Masters are transferred between owners without ever having to be lifted from their racks.

The freeport vaults are able to conjure ever more exchange value out of cultural artifacts that possess only abstract worth. The success of this alchemical process is dependent upon the freeport's international appeal and reach. For years the Geneva Free Ports and Warehouses Ltd., which operates the freeport, played up the idea that it was beyond the arm of government. The company boasted that it was a "genuine offshore base at the heart of Europe." Some accounts of the freeport even give the impression that it is a kind of autonomous statelet nestled in the securely unaligned body of the Swiss nation. It's an understandable mistake to make about a place that has for so long portrayed itself as untouchable. Reality intervened on September 13, 1995, when Swiss and Italian police raided room 23 on corridor 17. They discovered four thousand looted Italian antiquities. The Italian dealer Giacomo Medici was eventually convicted and jailed, at which point the Swiss decided it was time to regild the reputation of the freeport, although they took their time about it. They waited until 2009 to mandate full inventories and ownership details.

Far from frightening customers away, the new regime of scrupulous honesty has only increased the attraction of the freeport. There are huge profits to be made from perfectly legal transactions within its walls and a huge, perfectly legal demand for somewhere to store objects far away from tax annoyances at home. Clever accountancy has made criminal subterfuge an anachronism.

Speaking up for the public benefits of the Geneva Freeport, Jean-René Saillard, of the investors group British Fine Art Fund, explains, "Owners have every reason to lend generously. When works they possess are exposed by prestigious institutions, they naturally increase in value." But it's a rather backhanded defense. It suggests that art passes the freeport's doors only if it is on its way to accrue some additional cash value. It also implies that the role of galleries and art museums is to service the needs of tax-exempt storage and transaction sheds. The twenty-first century looks set to be the storage century. Art will occasionally be loaded up, trucked out, and made to perform in extra-storage public spaces. But as storage grows in influence, this periodic circulation is starting to be rationalized. The Geneva Freeport is already developing gallery space. In this way items can accrue value without the risk of loss or damage. Freeport warehouses are well positioned to emerge as the century's most significant exhibition spaces.

The Geneva Freeport has been a success story, but it may soon be overshadowed. Today many cities in the West, and increasingly in East Asia too, are dotted with warehouses offering shelf space for the world's glut of valuables. These urban bunkers are now part of the ordinary landscape: dreary concrete temples to the priceless and the unique. The largest growth in tax-free art storage is in East Asia. Singapore's art freeport began operating in 2010, and the twenty-acre Beijing Free Port of Culture at Beijing Capital International Airport opened in 2013.

The trajectory toward ever more storehousing challenges the idea that ours is an age of the virtual, an era in which real objects are increasingly irrelevant. While money flows by on computer screens, taking up no space at all, the storehouse culture has developed a power and life of its own, allowing and demanding more and more keeping, dusting, and inventory-taking. Such storage is only in its infancy, and it is not limited to art but includes all things of high value. Moreover, there does not seem to be any foreseeable limit on what can be amassed, since there is plenty of room under the earth or up in space. There may yet come a time when the valuable object is understood to have a life cycle, but we can't yet imagine that moment—when uniquely rare cars will be crushed, say, or precious art is fed to the flames. Until then, we will build more freeports, more storage, more shelves, more vaults for our treasure.

Bright Light, 4 Mures Street, Bucharest

44° 28' 04" N, 26° 02' 45" E

Number 4 Mures Street is an off-white single-story office building with plenty of large windows. It's a cheaply built, rather rundown 1960s construction and doesn't appear especially secure. It is fronted by modest metal railings. Behind this block, which lies on a dusty residential street in northeast Bucharest, there is a larger building painted in the same municipal shade, and behind that, some train tracks. Number 4 Mures Street looks like a place where officialdom is slowly churning and employees watch the clock, waiting to go home.

Between the end of 2003 and May 2006, 4 Mures Street had the CIA code name Bright Light. It was a secret interrogation and detention center, a so-called black site, and acted as a link in an international chain of covert detention and staging points used in the war against terror. For those detained here and for the rest of the world Bright Light did not exist—it was a non-place. We can call such sites non-places for two reasons: because these are spaces no one notices, and because they complete the bewilderment of the inmate— even when the hood is removed and the landscape glimpsed, it means nothing, it could be anywhere. One person who has made this connection is Bruce O'Neill, an anthropologist at Stanford University who has taken a special interest in the geography of "extraordinary rendition." In the journal *Ethnography,* he argues that secret detention facilities appear to "thrive in the non-places." They can grow and breed in these empty zones because no one cares about them: they are "generic and highly functional spaces that we pass through without establishing significant social or historical relations with them." It is places that "escape our serious attention or observation," O'Neill concludes, that provide "the ideal infrastructural site for a new kind of camp."

It is the forgettableness and unseen nature of 4 Mures Street, along with numerous windowless rooms elsewhere, that make it ideal for the covert extension of state power. The building, then as now, was owned by the Romanian National Registry for Classified Information. What a spokesman calls "media speculation that the building hosted a CIA prison" has been "categorically denied," and Romania's foreign affairs minister has confirmed that "no such activities took place on Romanian territory."

But the German media, specifically the *Süddeutsche Zeitung* newspaper and the ARD television network, have done more than speculate. Their interviews with former inmates and CIA agents pinpoint Bright Light in great detail. Journalists have been able to build up a plan of the complex and uncover the names of those who were kept there and what happened to them. What they found out was that the back building has a basement with six specially designed cells, each one built on springs. The idea appears to have been that a permanent sense of imbalance would disorient inmates, though ironically the cells also had an arrow painted on the floor to indicate the direction of Mecca.

Bright Light was used for particularly high-value inmates, including Khalid Sheikh Mohammed, who is accused of having planned the 9/11 attacks and was kept there before being transferred to the detention camp at Guantánamo Bay. Other prisoners included Abd al-Rahim al-Nashiri, who was accused of carrying out the attack on the US warship *Cole* in Yemen, and Abu Faraj al-Libi, Al Qaeda's third-in-command. Al-Libi identified Osama bin Laden's personal courier, a piece of information that eventually led to bin Laden himself.

During their short tours at 4 Mures Street, the CIA interrogators ate and slept in the compound, shut up with the inmates. A report from the International Committee of the

Red Cross, published in 2007, on the treatment of the kind of high-value detainees that passed through Bright Light, detailed the "fairly standardized" procedures that were intended to wear them down through a combination of humiliation and disorientation. In transit, the detainee "would be made to wear a diaper and dressed in a tracksuit," the report said. "Earphones would be placed over his ears, through which music would sometimes be played. He would be blindfolded with at least a cloth tied around the head and black goggles." Having reached an interrogation center like Bright Light, the "maintenance of the detainees in continuous solitary confinement and incommunicado detention" meant that many had no idea in which country they were. Kept in isolation, often for years, and without any information about their surroundings, the inmates were unable to provide a coherent account of their time in detention.

Even before the story of Bright Light broke, a Council of Europe report had concluded that "secret detention facilities run by the CIA did exist in Europe from 2003 to 2005, in particular in Poland and Romania." A couple of years later the *New York Times* established that the CIA's European headquarters in Frankfurt had overseen the construction of three detention facilities in Eastern Europe, "each built to house about a half-dozen detainees." One of these was in "a renovated building on a busy street in Bucharest."

It is not just the ordinariness of 4 Mures Street that deflects suspicion. Like many other government buildings in Romania, it was once used by the Securitate, Romania's secret police. Romanians have learned to leave such buildings well alone. Although the Securitate was dissolved in 1989, it had had four decades to mold Romanian society, and at its height it had 2 million people on its payroll, or 10 percent of the country's population. Romanians have vivid memories of the power of the security forces, who at one point had 180,000 forced laborers under their watch and a further 1.1 million political prisoners, held in more than 120 camps. While Romania's gulag days are over, they have left the country with both a physical and cultural infrastructure, a network of secret camps and public indifference to what might be going on in nondescript government buildings. Surrounded by an unseeing world and occupied by disoriented inmates, 4 Mures Street is a perfect example of a place beyond reach, a nonplace where old rules and old identities can be broken or forgotten.

Yet by the time we hear about such places, they are usually long gone. Bright Light closed down years before its story emerged. At the very moment that global attention homes in on one particular instance of secret detention, thousands of other camps and millions of other detainees sink deeper into the dark. In many countries, such "spaces of exception" continue to expand. In 2009 the Indian magazine *The Week* revealed that in India between 15 and 40 secret detention camps are operated by the intelligence services. In 2012 the Chinese government passed a new law giving the police the legal right to do something they had been doing all along: hold detainees at secret locations. It is estimated that today between 150,000 and 200,000 political prisoners in North Korea are held in dozens of secret camps. Windowless and unmapped, such non-places have become a resource for regimes that appear to have little else in common.

International Airspace

Humans are naturally terracentric. We're very interested in places on the earth but only vaguely aware of landscapes beneath the seas, and we rarely think about airspace as anything more than a void through which to travel. From the breath in our lungs to the sixty-two vertical miles that are stacked on top of us to the very edge of the atmosphere, air is a place both intimate and forever strange.

Whom does it belong to? Is there any that is utterly free? In any case, I seem to have inadvertently purchased some, since, in the United Kingdom, if you own any plot of land, you also own the earth below it and the space above. The ancient doctrine of property law, once enshrined in English common law, runs, *Cujus est solum, ejus est usque ad coelum et ad inferos:* "Whoever owns the soil, it is theirs all the way up to heaven and down to hell." Today that would open up some once unexpected issues, such as householders claiming toll fees from planes and satellites. So this ancient law has been whittled down. You are unlikely to be entitled to any more airspace than what is deemed necessary for the "use and enjoyment" of your plot. That still allows people to sell the development rights, or air rights, above their property. And if you want to build over someone else's property, you can buy someone else's unused air rights in order to do so. On a crowded island like Manhattan, where airspace is at a premium, air can be bought and sold for hundreds of dollars per square foot.

While property owners' rights have been deflated, sovereign airspace has ballooned. Indeed, it was the alarming increase in balloon flights at the start of the last century that first got statesmen interested in the topic. A conference in Paris in 1910 brought together eighteen countries worried by this new type of traffic to hammer out how airspace was to be controlled. The French wanted free skies, but the British liked the idea of full national sovereignty. No agreement was arrived at, but it wasn't long before the idea that the skies needed to be defended from foreign intrusion gained ground. The 1911 Aerial Navigation Act, passed by Parliament, allowed Britain to close its airspace to hostile flights. It was an early example of what was to become a century full of attempts to demarcate airspace as an extension of national territory.

Today national airspace extends up to twelve miles out over the sea, which still leaves plenty of the world's air unclaimed. Like the high seas, international airspace is open to all, but exercising that right, in a plane for example, is not at all straightforward. The trouble is that according to international law, you never really escape the place where your craft is registered. In other words, if your plane is registered in Norway, even when you are in mid-Pacific, flying between Fiji and Tahiti, you are still in Norway and you have to abide by Norwegian law. This precept also suggests that babies born on planes will sometimes be citizens of the country where the plane is registered and sometimes take their parents' citizenship (following different national laws). But it turns out that there might be competing claims on the infant. If you are born over the United States, in a foreign plane with foreign parents, you can still claim US citizenship.

Although it is hard to get access to it, the space above the high seas is free and remains unclaimed. There is also the possibility that if you go high enough, all claims to national sovereignty cease, since the traditional "up to heaven and down to hell" approach would mean that, as the earth rotates, great cones of sovereignty would arc their way across the galaxies. So there must be a limit, and this turns out to be a sore spot in legal circles, as lawyers have been arguing among themselves for decades about where the upper limit lies. Some say it ends where an aircraft can no longer get sufficient aerodynamic lift to fly; others make a claim for the zone that "permits the completion of one circuit by an orbiting vehicle." These are still vague limits, ranging from 43 miles to 99 miles above the earth. And even that distance isn't enough for some. The Bogotá Declaration of 1976 saw eight equatorial countries claim sovereignty to 22,300 miles. That is the spy zone, the zone of geostationary orbit, where a satellite's orbit is synchronized with the earth's rotation, enabling it to effectively sit still over a country. The Bogotá group's claim turned out to be unpopular, though, partly because it violated the idea that space is our "common heritage."

The arguments continue, but in the meantime we can safely assume that high above nations there is plenty of space, between the lower atmosphere and outer space, that is beyond any country's jurisdiction. Combined with all that airspace over the high seas, it would appear that the bulk of the world's thin rim of atmosphere is outside of national or private control.

So what shall we do with it, if we treat the air above us as somewhere to go, possibly even to live? The idea of airborne cities has been intriguing architects for some while, and a number have been planned by the profession's blue-sky thinkers. Some of the first schemes were set out by the counter-cultural architects Archigram, whose Instant City envisaged an entire city hoisted aloft by balloons and helicopters that would then move it anywhere its citizens felt like going. Such a place could easily escape national airspace. It's an idea that has been extended by others, most recently by focusing less on transporting cities and more on the business of keeping aloft. Utopian architect Leah Beeferman envisages a kind of scattered republic: "The helicopters themselves could be liberated to form their own city—an airborne utopia, endlessly aloft, wandering through the planetary atmosphere."

For Beeferman, this "helicopter archipelago, or flying island-chain," would be "an escape hatch from traditional, nation-state sovereignty." It would roam across international airspace: the "archipelago would be impossible to map. Atlas-makers and manufacturers of globes will simply include a pack of removable stickers, featuring small clouds of helicopters, to approximate the country's location." It is somehow fitting that the idea of free airspace should deposit us on cloud nine. Although a strange and alien environment, the air is also the natural place for far-fetched bids for human happiness.

Gutterspace

In the early 1970s the conceptual artist Gordon Matta-Clark began buying up shards of land between buildings in New York City, called gutterspaces. Most are just a few feet wide, though they are often hundreds of feet long. They are the useless remnants of the planning process auctioned off by the city authorities. Matta-Clark bought fifteen (fourteen in Queens and one on Staten Island), and over the next few years he took photos and collected the plot plans, deeds, and all the bureaucratic documentation he could get hold of associated with each of his new properties.

Most of these gutterspaces are alleyways, sunk in shadow; lurking between buildings, they have a furtive, melancholy look. It seems that Matta-Clark was drawn to them precisely because they were advertised as "inaccessible." In an interview he explained that he liked the paradox of owning spaces that "wouldn't be seen and certainly not occupied." But Lot 15, Jamaica Curb, is different: it is a long rim of sidewalk that sits in broad sunlight, an accessible, apparently public space and a flagrant bit of legal and commercial oddness. Matta-Clark's twenty-four photos of it form an eleven-and-a-half-foot-long collage.

Matta-Clark died in 1978 at the age of thirty-five, leaving his project, which he called "Fake Estates," unfinished. It was already getting irksome, largely because of all the separate property taxes he had to pay. The lots, which cost him as little as $25, each incurred a tax of $7 or $8 a year. Although after he died the land reverted back to city ownership, Matta-Clark's untidy piles of papers were assembled and put in order by his widow, Jane Crawford, and, over the past couple of decades, exhibitions, books, and bus tours have been constructed around them. "Fake Estates" has become a touchstone for a new generation of psychogeographical artists.

Anyone who has watched nervously as their next-door neighbor put up a fence knows just how important a few inches of land can be; after all, not many of us would lean out the window and politely call, "No, really, don't worry, just put it anywhere." For all the disconcerting beauty of "Fake Estates," Matta-Clark thought he was providing a *reductio ad absurdum* critique of our obsession with private property. In her study of his work, *Object to Be Destroyed,* Pamela Lee explains his motives by reference to Karl Marx's claim that "private property has made us so stupid and narrow-minded that an object is only ours when we have it, when it exists as capital for us." Matta-Clark's useless lots are supposed to destabilize our sense of the rationality of land ownership—and they do, in a way. But they also work in the opposite direction, because they remind us how powerful the idea of owning land really is. It is because this desire is so commonplace that "Fake Estates" is so oddly touching. Matta-Clark's gutterspaces speak to a widespread yearning for a piece of the earth—no matter how small—that we can call our own.

Matta-Clark's collection of pieces of land also taps into an aesthetic of ordering and labeling that we are all familiar with. In fact, at about the same time as Matta-Clark's conceptual activities, another New Yorker, a hardware store owner named Jack Gasnick—already a minor celebrity for his "cellar fishing" (he claimed to have caught a three-pound carp in a stream running through his basement) and for having the world's third-oldest working lightbulb, at the back of his store (it was first turned on 1912)—was doing something very similar and without a whisper of political intent. For Gasnick, buying up gutterspaces was a hobby. Interviewed in 1994 by Constance Hays of the *New York Times,* he explained, "It's like collecting stamps," adding that "once you've got the fever, you've got the fever . . . I wanted the unwanted."

Gasnick's set of gutterspaces eventually came to twenty-eight, bought for between $50 and $250, and his prize item was some land behind Louis Armstrong's house in Corona, Queens. But Jack had many strips and squares, including part of an African-American cemetery and one with an apple tree on it. While Gasnick's relationship to these bits of land was that of a collector, he also thought of them in aspirational terms. "This jump of mine from flowerpot to apple tree," he said, "bears witness to the fact that it doesn't cost much for an apartment-living guy to get a share of the good environment." The care he took of his apple tree, not to mention the oak in another plot, reflects Gasnick's real affection for his micro-plots. "Once he acquired a plot of land," Constance Hays wrote, "Mr. Gasnick spent weekends driving out to visit it and clean it up." Hays describes how Gasnick "kept visiting his properties even as the neighborhoods around them changed dramatically, keeping his land clear of litter, whether empty coffee cups or abandoned cars." In the late 1970s Gasnick began to feel overwhelmed by the upkeep, and he either mislaid lots ("some I just forgot") or sold them off, often for little more than he paid for them. Other lots were given to community gardening organizations. Now in his nineties, he has hung on to one last lot, a beloved picnic spot that offers a harbor view from Staten Island.

Snippets of land still come up for sale. Claudio Manicone, who works for Broward County, Florida, has been trying to sell a sidewalk, an alleyway, and a river. He sees it as tidying up: "Like in anything you do, you've got leftover pieces. Just like when you build something out of wood, you've got wood left over." The companies that sell these parcels occasionally encounter buyers with ulterior motives, such as the man in Palm Beach who bought a tiny section of municipal canal thinking that this would give him control of the water supply (he soon found out that it didn't). But mostly people are interested in these places for far less rational reasons.

Gordon Matta-Clark brought his remnants out of obscurity and onto the map. It was always an ironic achievement, since they remained "fake estates," real but worthless. But even so, it is easy to see why people would want them, especially in cities where most of us feel lucky if we have a yard large enough to sit down in. And I think I understand why Jack Gasnick felt good about his collection. Since the rich take such delight in their spreading acres, is it so peculiar that ordinary people might get some pleasure from owning a few feet of grass? And if people looked bemused, you could always explain that your miniature estate is a critique of capitalism. They would surely understand.

Bountiful

54° 20' 00" N, 81° 47' 15" E

The relationship between place and well-being seems to be hard-wired into the human brain. Making a better life for oneself suggests going to a better place, and it comes as no surprise that creating a new kind of place is central to the efforts of those who want to flee industrial civilization and fashion a perfect society. Once associated with hippie communes in the 1960s, this utopian impulse has spread and diversified, and today there are a huge variety of "intentional communities," with the fastest expansion among eco-friendly and off-grid settlements. There are thousands throughout the English-speaking world, but my example comes from the banks of the Ob River, 1,750 miles east of Moscow and 65 miles south-west of Novosibirsk, Siberia's largest city.

The desire to escape urban life and build utopia is not new in Russia, but over the past few decades, in the wake of the collapse of the USSR and the social disintegration that has taken hold in many towns and cities, a new generation is packing its bags and heading into the forest, like the sixteen families who now live two miles down a dirt road in the village of Bountiful.

The fact that we take the bond between place and utopia for granted is a paradox. Utopia is Thomas More's Greek neologism for "no-place," a term meant to contrast with real places, which are frustratingly but inevitably full of different histories, ideas, and people. Utopia is an idea that implies that a utopian place could never work. Yet the intention to start such places is not uncommon; many ordinary towns and suburbs started life as ideal communities. There is a creative friction between such intentions and the realities of place-making that undermines utopian purism but also sparks new utopian projects.

Bountiful is part of the new phenomenon being labeled Russia's Green Exodus. Hundreds of eco-villages have been founded in the forests, often by professional, educated people who are turning their backs on what they see as the corrupt and corrupting nature of modern Russia. Many hark back to the anarcho-Christian and Tolstoyan communities that were formed in the first decades of the last century, only to be dismantled by Soviet collectivization, and to the small agricultural cooperatives championed by another victim of the USSR, Alexander Chayanov. Although the Green Exodus is too diverse for easy generalizations, it is often marked by a semi-mystical yearning for a purer, kinder, and more authentic Russia.

Bountiful is part of an eco-spiritual sect called the Anastasia Movement, based around a series of nine books written by Vladimir Megre, in which he claims he met a beautiful young woman called Anastasia on the banks of the river Ob in 1994. He recounts that her parents had died shortly after she was born and that she "has ever since fended for herself, watched over only by her grandfather, great-grandfather and a variety of 'wild' animals." Anastasia revealed to Megre a philosophy of "eco-culture where every person is fulfilling their role as a Divine Co-Creator" and instructed him that "every person has the right to a small parcel of land to grow their own food, build their own house, and raise their family, without taxes."

It turned out to be a timely message, because the Russian government was keen to privatize and diversify land ownership. Growing your own food and tending your own patch of ground is extremely popular in Russia—it was estimated in 1999 that 71 percent of the country's population already owned a plot and were cultivating it. In 2003, the same year that saw the founding of Bountiful, the Private Garden Plot Act allowed Russian citizens to claim free land of between one and three hectares. The Anastasians make much of the Russian prime minister Dmitry Medvedev's claim that because of "the scale of a country like ours, with such huge areas, there is no point everyone concentrating in cities," and that it would be "more useful for our health, and for the country, to disperse."

The Anastasian message also chimes with the rise of cultural conservatism in Russia. The movement's emphasis on traditional family values, and reverence for Russian crafts and home cooking, suit the temper of the times. Each homestead in Bountiful, of no less than one hectare, can only be inherited and never sold. Unlike some of the more hippie-ish villages spawned by the Green Exodus, this is one place that is consciously and nostalgically finding its utopia in the past.

Yet part of this nostalgia is to reclaim a history of collectivism and mutual care. Households help each other as well as new arrivals. The family of former physicist Valery Popov shows newcomers how to build their log cabin. Another family, the Nadezhdins, former dentistry professionals, are the village bakers. A music teacher named Klavdiya Ivanova turns out traditional Russian clothes. These local skills are highlighted on Bountiful's well-developed website, alongside cheerful stories and tips about how to return to a more wholesome and Russian way of life. One of the leading residents of Bountiful, Dmitry Ivanov, a former navy officer who helps install the cabins' stoves, explains, "The motherland is what teaches us to live in harmony." For all the New Age rhetoric, this is a place that flows with a patriotic desire for reattachment to the real Russia.

Imparting traditional Russian values within a spiritual, environmentalist context has proved an attractive formula. The Anastasia Movement now claims more than 100,000 registered activists and 85 villages, some much larger than Bountiful, spread across Russia (there are 4 Anastasian villages in the Novosibirsk region alone). But the fact that place and utopia are never an easy mix is also being continually proved. For one thing, the Anastasian philosophy is far from universally revered. Indeed, according to Ivanov "it is not so important" and Bountiful is doing its own thing: "It is more important that we are choosing the path that we are going, whether or not it is Megre's."

A bit of classic geography can help us understand how utopian places maintain their cohesion. Geographers like to identify the "push and pull" factors when working out why any settlement gets going. Bountiful has strong attractors: leaders and an ideology that are drawing people in. But the powerful forces that are driving people away from conventional places seem to be just as important. Listening to interviews with village members, one repeatedly hears the same story about what has propelled them so far away from the city and "the system." Talking to a visiting freelance reporter, Ivanov explains, "All my life, I've been a part of the system. At school, as a university student, then as a faithful officer." But the system failed him: "the system fell apart before my eyes, destroyed by traders, by stealers, by outrageously corrupt managers." Bountiful is dedicated, says one young mother, "to our future children so they can be more 'real' than we are." Olga Kumani, a resident of a nearby eco-commune at Askat, describes how she left her job as a crime reporter in Novosibirsk in 2002: "I could not breathe in the city; the state system choked me." But it didn't work out. "The commune leaders just wanted to control our money and exploit us for work." So the push factors kept pushing, driving Olga to move to an even remoter part of this vast region.

What holds utopia in place is not just a vision of a perfect place but the experience of living in a bad one. Ironically, the bad places were often former ideal places, their failure provoking people to seek out better alternatives. However, Bountiful also shows that the pursuit of utopia can quickly become a workable project: as ideological purism cedes ground to ordinary needs, it can offer both tangible benefits to individuals and concrete examples of social change. Today the "no-place" of utopia has thousands of outposts. Some are remnants of past hopes while others are newly founded, but all illustrate humanity's powerful and paradoxical hunger for escape and homecoming.

Mount Athos

40° 09′ 32″ N, 24° 19′ 42″ E

The Holy Mountain, Mount Athos, is a fifty-kilometer-long peninsula that kicks out into the Aegean Sea. Along its coast are twenty Greek Orthodox monasteries, steeply walled and turreted. Most of them were founded over a thousand years ago, their thick defenses and lofty towers bolstered over the centuries as protection from pirates. Athos also contains the medieval town of Kayres, the village of Daphne, and many chapels and ancient ruins. It is a wild and rugged landscape, accessible only by boat, spined with mountains that rise to two thousand meters or more at the southern end.

It's not off the map for me, but it might be for you. Athos is an extreme example of a place defined by exclusion. Women are banned; even curious female sightseers are supposed to stay at least five hundred meters offshore. If they reach land, they are subject to a period of imprisonment ranging from two months to a year. Not only are women banned but so too are all female animals. One of the few exceptions is female cats, which, according to the monks, were "provided" to them by the divine providence of the Virgin in order to control vermin. But cats aside, permission to visit is restricted to adult men and "young males accompanied by their fathers."

The desire for men-only religious places may appear anachronistic, yet the story of Athos shows that it is remarkably resilient. According to legend, Athos was given by God as a holy garden to the Virgin Mary. On her way to visit Lazarus in Cyprus, storms swept Mary and her companion, John the Evangelist, onto the peninsula's east coast. They landed at a spot near a pagan temple dedicated to Apollo. Today it is the site of the monastery of Iviron. It is said the "pagan idols" cried out to the local people to come down and greet Mary, which they did, abandoning their old ways and converting to the new faith. Struck by the beauty of the area, Mary prayed to God to have it given to her. God spoke to her: "Let this place be your lot, your garden and your paradise, as well as a salvation, a haven for those who seek salvation."

Mount Athos is dedicated to the Virgin, and the bulk of its many icons are images of her, but it remains a male sanctum. When the legality of the ban on women is questioned, it is argued that its 335 square kilometers must be understood as one big monastery. "If one views each of the 20 monasteries of Mount Athos as a single entity," explained one of the Holy Mountain's secular champions, Austrian politician Walter Schwimmer, at a recent international conference dedicated to Mount Athos, "the ban of women from a male monastery is nothing extraordinary but a rule that is commonly accepted." Schwimmer's argument relies on and thereby highlights the fact that spatial exclusion remains a widely accepted facet of religious life. Occasionally its biggest impact is on nonbelievers. Two of the most visited places in the world, Mecca and the center of Medina, are also two of the most inaccessible: non-Muslims are forbidden. Mormon and many Hindu temples are also off-limits to nonbelievers, but such attention to faith is the exception rather than the rule. Usually it's not faith that matters when getting into religious places but gender. With the exception of some reformed Christian and Jewish denominations, the world's religions exhibit a deep sense of anxiety about the presence of women. Until recently women were forbidden to enter the sanctuary in a Catholic church, and the Muslim and Hindu tradition of purdah keeps millions of women trapped in their houses or peeping out at the world from behind the "protection" of a veil. In some of the remoter Hindu villages of Nepal the practice of *chaupadi* survives. This tradition dictates that women must not enter their own homes for up to seven nights during menstruation. Instead, they must live and sleep outside, in huts, in caves, or in the open.

From a traditional religious perspective men and women living side by side in towns and villages is a source of endless problems. These can be solved only by choreographing rituals of separation. Mount Athos is free from such headaches. It is a utopian space in which the celibate holy man's wish—to live without distraction and temptation—is finally realized. It's the best that earth has to offer until the day of resurrection, when men can finally shed their mortal bodies.

From its famed "six thousand beards" the number of monks on Athos has dwindled to about two thousand. They constitute a self-governing community whose political autonomy is enshrined in Greek law: the Greek constitution recognizes Athos as "a self-governed part of the Greek State, whose sovereignty thereon shall remain intact." The only bishop with authority over Athos is the "Ecumenical Patriarch in Constantinople–New Rome" (which the rest of the world calls Istanbul), where in 1046 Emperor Constantine Monomachos sanctified the exclusion of women from Athos.

This prohibition of women has its own legal name, the Avaton. As a piece of legislation it must be judged, on its own terms, a success. Despite its long history and vaunted beauty, the number of women who are known to have entered Athos is very small. There was Helena of Bulgaria, who came here to escape the plague in the fourteenth century. But she may not count, as her feet never touched the ground. In deference to local mores she was lofted around in a hand-held carriage during her entire stay. Athos met a firmer pair of feet when Maryse Choisy, a French psychoanalyst and onetime patient of Freud, decided to pay a visit. She put on a large false mustache and dressed as a male servant. She also claimed to have undergone a bilateral radical mastectomy, what she called her "Amazonian." Her commitment paid off and she spent a month on Athos. In her book *Un Mois Chez les Hommes* (1929), Choisy records the following interesting clarification from a monk at Vatopedi monastery on the prohibition of hens, which are banned as part of the general prohibition of female animals: "We must draw the line somewhere," he explained. "The day we possessed a hen, some brothers would argue that we should also accept a she-cat, a ewe (a useful animal) or even a she-ass. And there is but a step from a she-ass to a woman." The monk's list of banned animals suggests that the admission of female cats to Athos is a relatively recent concession. Enraged by the misogyny she encountered, Choisy delighted in exposing the Holy Mountain, depicting the monks as lazy, slow-witted, and racked by homoerotic desire. Her account is mocking and salacious, and its sexual content has been dismissed by some as vindictive fakery. Single-sex communities are ready fodder for titillation. But it's a leering curiosity that points to a real paradox: sex may be repudiated in such places, but it is also their organizing principle and, hence, their obsession.

Athos will probably always be plagued by occasional female incursions. Yet far from making the monks doubt their territorial gender claim, invasion and ridicule only seem to strengthen their view that they are the defenders of a sacred heritage. The Avaton is just one of many ways in which Athos is proudly out of step with the modern world. It permits foreign visitors on sufferance. The monks issue just ten visitor permits a day to the non-Orthodox, but up to one hundred for "Greeks and Orthodox." When they land at Athos, visitors have to go back in time, literally. The Greek Orthodox Church adopted the Gregorian calendar in 1924 but not on Athos. Here the monks still use the ancient Julian calendar (a practice also maintained by small Old Calendarist sects in Greece, Romania, Bulgaria, and the United States). As a result, Athos is thirteen days behind the rest of the world.

The stubborn archaisms of Athos and the beauty of its landscape make many feel protective toward it, including Prince Charles, who is a regular visitor. Yet there is something deeply unattractive about the sophistry that sustains the Avaton. Today its defenders resort to the language of respect for cultural difference. Walter Schwimmer argues, "Somebody who demands the end of the ban of women on Mount Athos simply lacks respect for the way of life the monks of Mount Athos have chosen." Schwimmer throws us an apparently rhe-

torical question: "Can such lack of respect for the others, vio-
lating their human dignity, be the basis of a 'human right'?"
It's a slippery argument and makes it hard to warm to Athos.
Such a defense could be used to deny any and all human
rights merely on the basis that they impinge on someone else's
choices. The "respect my choice to discriminate" defense also
reminds us that, when considered alongside the world's mani-
fold examples of religiously sanctioned female-free places, the
Holy Mountain looks like an extreme case of a general trend
rather than a charming exception.

Ranch of Sprouts: Brotas Quilombo

23° 00' 59" S, 46° 51' 31" W

Ranch of Sprouts is one of the old names for Brotas Quilombo,
an Afro-Brazilian township. Quilombo is the generic name
given to the free territories that were established by runaway
slaves in Brazil. The settlements came in all shapes and sizes,
but the most famous was Palmares. Established in 1600 on
the northeast coast of Brazil, Palmares was a republic of ex-
slaves and is said to have been about the size of Portugal. It
held out for eighty-nine years, far longer than most quilombos.
Today there may be two thousand suburbs and villages that
have quilombo roots. The Palmares Foundation, which is attached to the Brazilian Ministry of Culture, recognizes 1,408,
spread out across all but three of Brazil's twenty-seven states.

The quilombos were forgotten for most of the past century. Unmarked and unrecognized, they hung on as scruffy enclaves, absorbed and surrounded by modern municipalities that expropriated their land or did their best to ignore them. But the Brazilian constitution of 1988 changed all that when it recognized the legitimacy of quilombo land. It was a momentous shift, for the constitution declared: "The definitive property rights of remnants of quilombos that have been occupying the same lands are hereby recognized, and the state shall grant them title to such lands." In 1995 a year of national celebration was announced to commemorate the death of Zumbi, the last leader of Palmares, who was officially titled a "hero of the Brazilian nation." In this atmosphere of acceptance, many communities have outed themselves as quilombos in "festivals of self-definition," usually held on November 20, Brazil's annual "day of black consciousness."

The quilombos have become the centerpieces of a new confidence and pride among Afro-Brazilians. Their acknowledgment has also opened up interesting questions about the centrality of "free places" in the struggle against slavery and for black identity. Escape is not just about running away; it's about having somewhere to go, about setting down roots in a different kind of place. If free places cannot be sustained, then escape becomes impossible and resistance slowly dies. The story of the quilombos drives home these simple truths, but it also throws up more complicated issues. After all, if quilombos are communities of escape, what is the point of them after the abolition of slavery and in a world in which Afro-Brazilians are just another ethnic group in a multiracial society? By referring to the quilombos as "remnants," the constitution was making a point: it was recognizing their history but also dispatching them to the past. When does a place stop being a quilombo? When does it stop being defined by its past? The answer has emerged over the past couple of decades, as the quilombo movement has evolved, and for the time being at least, it appears to be "never."

Brotas Quilombo is being tested by all these questions. It is home to about thirty families and lies on the edge of Itatiba, a small town seventy-six kilometers from the booming megacity of São Paulo. The houses are built with concrete blocks and asbestos tiles and are spread over an overgrown suburb of dirt streets and tropical woodland. This modest place was once just part of a much larger escape zone that drew in runaways from far and wide. However, like many other quilombos, a small plot of land was eventually bought up by two ex–farm slaves, Emília Gomes de Lima and Isaac de Lima, the original novice or seedling ranchers, hence the name "ranching sprouts." Oral history testimony cites Isaac de Lima as hoping that "everyone that has my blood will have a place to live." Many of the villagers root their ancestry in this first tenured couple, and they are the great-grandparents of its oldest inhabitant and local matriarch, Ana Teresa Barbosa da Costa, whom most people in Brotas Quilombo call Aunty.

The ownership of Brotas Quilombo has been in jeopardy many times over the past hundred years. Unaffordable property taxes, lack of official recognition, and plans to turn the whole site into a hospital waste dump nearly destroyed it several times. Until recently it was off the beaten track and at the wrong end of town. Few people visited apart from those going to worship at the quilombo's popular Umbanda religious or cult house. Interviews with the people who live in Brotas

Quilombo have consistently showed a powerful desire not to let the past slip through their fingers. "Today, eight generations later, most of the residents of the quilombo are of mixed race," a local resident, Paulo Sergio Marciano, told a BBC reporter. "But our priority is the recovery of our traditions, of the connection between Brazil and Africa." Today this nostalgic passion is finally reaping rewards, and Brotas Quilombo has been recognized by the state of São Paulo as an urban quilombo, an achievement that has brought with it a cluster of official and academic reports on its past and future.

Coming out of the shadows has also brought concrete challenges, some of which concern property law. Like most quilombos, the ownership deeds for Brotas Quilombo were far from complete. Entitlement to the land has had to be proven through many different sources, including oral histories and, literally, dug-up artifacts. Chains and iron balls have been uncovered and passed on to the authorities as well as a stone figure of a woman from an African tribe. These items matter, for they are used as evidence of authenticity and ownership. This was especially important in the early 2000s when residents were faced with a construction company that had started to build condominiums on their land. In 2003 a government official claimed that the state was "not in a position to say whether or not there has been an invasion of this property, because there are no obvious borders." Since then much has changed. The borders have been confirmed and new investment has brought street lighting and a heritage center to Brotas Quilombo. Indeed, there has been something of a revolution of attitudes. From a marginalized backwater it has become a fashionable place to visit, especially for a more socially conscious generation. It now hosts interracial festivals and history events that attract people from all over the district.

The "quilombolization" of Brazil is seeing once hidden communities step into the light. Even settlements that make no claim to being founded by escaped slaves are seeking and now gaining recognition as quilombos on the basis that they have a largely black population. Interviewed by a fellow anthropologist, Alfredo Wagner Berno de Almeida, who works with quilombos across the country, warns against casting them as a frozen heritage. "The quilombo is not the sphinx, it is not a pyramid," he says. "They are not monuments, they aren't part of the artistic patrimony. They are part of the productive life of the country." Others, like Onyx Lorenzoni, a federal deputy, complain that the march of the quilombos "is dividing Brazil into nations of color." Is Brotas Quilombo turning from an outpost of Afro-Brazilian culture to a subsidized black time machine? It doesn't seem so. Locals of all different shades and backgrounds are coming to Brotas Quilombo and getting involved. Its rediscovery is adding something unique and important to the wider town.

Quilombos are not remnants of something gone but places that look to the past to define their present. It's something that all living places do. It can sometimes mean that they appear to be more interested in preserving traditions than inventing new ones. But that is a risk worth taking—indeed, it is a risk that has to be taken if places are to be communities, something more than just spaces of temporary individual habitation. Without the binding presence of the past, places are emptied of a meaningful future.

FARC-controlled Colombia

In an increasingly surveilled and homogenized world, some may consider insurgent places an anachronism. If so, the Fuerzas Armadas Revolucionarias de Colombia (FARC), the Revolutionary Armed Forces of Colombia, hasn't heard the news. In the early 2000s they controlled around 40 percent of the country. Over the past decade their zone of influence has diminished to a little less than 30 percent, but that still amounts to a great deal of jungle.

FARC-controlled Colombia is unique territory, not only in terms of its scale and longevity as an insurgent place but also because FARC's capacity and desire to hold on to this place seems to run counter to global trends. It's not that heavily armed militants have gone out of fashion but rather that we have grown used to a different breed. Islamist groups move around like ghosts, quietly slipping between rented rooms and anarchic nations. Their only hope of territorial control arises from the connivance of states that find them useful, or failed states that can't get in their way. Somalia, Afghanistan, Sudan, South Yemen, Mali, and Pakistan have all fallen into one or both of these categories. Al Qaeda, Arabic for "the base," found such a base in Taliban-controlled Afghanistan. Away from such protected safe spots, militants have had to morph into unplaceable networks. The results have changed all our lives, so it's not surprising that our ideas about the geography of armed insurrection have been shaped by religious fighters' spectral maneuvers.

As a result, a very different kind of rebellion, with a differ-

ent relationship to place, has been overlooked or cast as yes-
terday's news. Yet armed rebels with revolutionary socialist
ideologies have not gone away. They include Maoists in Nepal, who in the 2000s controlled about 80 percent of the country, and Naxalite Maoists in India, who currently have operational control over large swaths of remote forest. Another
example is the socialist nationalists of the Kurdistan Workers' Party, the PKK, along with their political offshoots. They
are based in the Qandil Mountains of northern Iraq, but
their sphere of control extends deep into northern Syria and
Iran.

These leftist revolutionaries have often shown scant regard for human life. However, they are bottom-up move-
ments, peasant-based and with many female cadres, which,
geographically, makes a great difference. It means they are
rooted in place and can and want to hang on to the territory
they consider their home turf. The contrast with Islamist terrorism, which is a top-down, entirely male-dominated, geographically restless phenomenon, is stark.

The FARC has a real stake in the place from which it
comes. It was founded in April 1966 by communist farming
communities that had been fighting a war of "mass self-de-
fense" against government forces for nearly two decades. The
FARC developed in rural areas that had a long tradition of
fierce autonomy and distrust of the central government in the
capital, Bogotá. By the 1980s FARC control had spread well
beyond its core areas. Militarily the guerrillas were pushing
forward on more than eighteen fronts, proudly adding the
term "Army of the People" to their name. They were "operating as a de facto government," says FARC expert Gary Leech,

"for rural communities across vast stretches of countryside where the state had never established a presence."

Over the past twenty years the FARC has come to embrace a more flexible, more nationalistic, "Bolivarian socialism." It has left many of the strictures of Marxism-Leninism for its sometime comrades, sometime enemies, in a smaller outfit called the National Liberation Army (whose zone of influence is in the mountainous northwest of the country). But to understand how and why the FARC continues to place such store in occupying the jungles and remote villages of Colombia, we must look to some classic revolutionary works. *Guerrilla Warfare,* published in 1961 and written by Ernesto "Che" Guevara, is probably the most important. Establishing and keeping inaccessible territory is the key to Guevara's lesson plan. A "guerrilla band will here be able to dig in," he says. "Aircraft cannot see anything and cease to operate; tanks and cannons cannot do much owing to the difficulties of advancing in these zones." Building up defensive space also allows a revolutionary infrastructure to be put in place, "where small industries may be installed, as well as hospitals, centers for education and training, storage facilities and organs of propaganda."

But these technical details only scratch the surface of Guevara's ambitions. His real motivations were political. Rooting an armed group in the countryside allows the revolution to go to and come from "the people." Building up what another guerrilla leader with an obsession for holding territory, Mao Zedong, called "the base of the people" is the means and the goal. The idea is that "guerrilla warfare basically derives from the masses and is supported by them," Mao said. "It can

neither exist nor flourish if it separates itself from their sympathies and co-operation." Given the ruthless nature of the
FARC, which gains much of its income from kidnapping and
taxing drug traffickers, it may sound like an unlikely conceit.
But it provides the guerrillas with what they consider to be
a tried and tested basis for their own commitment to redraw
the map of the nation into zones of insurgency and imperialist
power.

So much for the theory. The FARC has not shown itself adept at maintaining the "sympathies and co-operation" of the Colombian people. One 2001 Gallup poll showed
that less than 3 percent of Colombians had a favorable opin-
ion of the FARC. Many blame its fighters for a war that has
killed around 250,000 people and caused millions to flee their
homes. It is a war that looks both endless and pointless. Although the FARC's determination to control territory has
been successful, the group never seems to have known what
to do next. Che and Mao would have pushed on, because
they saw the mountains and jungles as a springboard. But the
FARC guerrillas aren't just based in the countryside; they are
entombed in it.

What keeps them there is partly sentiment but also the
brutal fact that these are the only areas where their message
continues to have any kind of appeal. Writing for a Colombian political journal, Alfredo Rangel, a onetime defense ministry official turned Bogotá-based private security worker, explains that while "their national banners are invisible or not
credible, their local, armed patronage and their ability to take
advantage of rural youth unemployment allows them to establish pockets of support in many regions."

The result is an endless series of halfhearted advances and

speedy retreats—over recent years more of the latter than the former. The longer this has gone on, the more FARC's investment in carving out territory has shaped the mindset of all the other combatants on the field. The Colombian army also defines its mission as seizing, holding, and defending territory. So too do the anticommunist paramilitaries of the Autodefensas Unidas de Colombia, which in the mid-1990s declared that its intention was to "reconquer" land "colonized" by the guerrillas, "because it is there that the subversion has succeeded in creating parallel government."

The stalemate looked as if it might be coming to an end in 1998, when the government handed over a huge swath of territory to appease the FARC. It was hoped that this gift, an area about the size of Switzerland, would provide the kind of gesture of goodwill that would kick-start meaningful peace talks. Although the détente lasted only a few years—the FARC seemingly unable to get beyond warfare as their raison d'être—there have been growing calls in recent years for the idea to be given another chance. Spain and France have proposed that an international demilitarized zone, without the armed presence of either the guerrillas or the state, might be a better way forward. It is a territorial solution to what is a territorial dispute.

Whatever the success or failure of this latest initiative, it has shown beyond doubt that the FARC, and organizations like it, set enormous store in possessing place. It is the fight for it that defines them. What differentiates them from terrorists and makes them genuine revolutionaries is that they are bound to place. Most modern terrorism, by contrast, is placeless; it thrives in an uprooted world. The problem for the FARC guerrillas is that their geographical ambition has come to outweigh all others. They hang on to place so tightly that they have squeezed the life from it.

Hobyo

5° 20' 59" N, 48° 31' 36" E

Hobyo has seen better days. An ancient town on the western coast of the Horn of Africa, it prospered for many centuries thanks to the Indian Ocean's busy shipping lanes. A hundred years ago it was the capital of a small sultanate and a lively commercial center, drawing in traders in precious metals and pearls. Today it is a pirate town that is avoided by the rest of the world. But it's at night when Hobyo really disappears from the map. For in a town where a single hijacked vessel can make more than $9 million, it's surprising to learn that no one has invested in an electricity generator. Nighttime satellite images show Hobyo as an inky nothing. It's a place that sees a lot of money but is dirt poor, since both the economy and the identity of the town have been hollowed out by brigandry.

In the 2000s there were plenty of pirate towns up and down this coast, but Hobyo was one of the most secure. When they were beaten back by Islamists in the south or the marine police in the north, the pirates came to Hobyo, sometimes bringing their booty with them. As a result, Hobyo is as good an example as any of a "feral city." It's a term that is used in military circles to describe regions that have no effective government but sustain an internationally networked criminal economy. Feral cities are the ragged end of spaces of exception: they are not the product of governments or ideologies but show what happens when such structures fall away. According to Richard Norton, writing in the *Naval War College Review*, feral cities have "lost the ability to maintain the rule of law," yet they remain "a functioning actor in the greater international system." Norton throws his net wide, arguing that Mexico City, São Paulo, and Johannesburg have entrenched feral characteristics and may be well on their way to becoming feral cities.

These cities may be partly feral, but Hobyo is almost entirely so. Somalia has had no effective central government since its 1991 civil war, and although it lies within the Somali province of Galmudug and has its own mayor, Hobyo has been in the hands of pirates for over a decade. Its remote coastal location makes it one of their prize possessions. Planted a mile off Hobyo's coast lies the pirates' hoard. "This one is bigger than Hobyo," gloated one young pirate to a visiting French journalist, Jean-Marc Mojon, pointing offshore to a hijacked Korean supertanker that was soon to net his comrades millions of dollars. Meanwhile, pirating has spawned a secondary industry, with small fishing boats taking supplies from Hobyo to the stolen vessels. Satellite images show that the town is dotted with heavily walled pirate compounds, their courtyards lined with vehicles. The town's most prominent feature is a telecommunications tower, which is used by the pirates to communicate with hijacked ships anchored off the coast.

In recent years the total paid over to Somali pirates has been between $150 million and $200 million per annum. The size of this figure comes more clearly into relief when put alongside Somalia's GDP per capita, which is $300. All of which makes you wonder where all that ransom cash goes. Not to Hobyo, that much is certain. It ends up abroad or someplace far inland. Pirates aside, Hobyo is a dusty, low-rise, crumbling town fighting a losing battle against the encroaching desert. Other interviews by Jean-Marc Mojon with the residents paint a bleak picture of a place emptied of hope or purpose. A local elder complains that "we have no schools, no farming, no fishing—it's ground zero here." The idea that piracy should be reaping benefits for Hobyo doesn't appear to have occurred to him. Instead, he worries about the desert: "our most pressing concern is the sand, the city is disappearing, we are being buried alive and can't resist."

Hobyo's pirate bosses offer what has become a standard justification for their trade: foreign vessels came into our waters and stole all our fish, forcing us into piracy. But most serious analysts don't buy the idea that the pirates would much rather be fishing. While it is clear that illegal foreign fishing in the early 1990s did first provoke local fishermen to arm themselves and defend their waters, thus forming the nucleus of the pirate fleets, evidence collected by the Norwegian expert Stig Jarle Hansen suggests that fish stocks remain viable. The scale of the switch from fishing to piracy is better understood as an economic and social choice. In a poverty-stricken country the quick and vast financial rewards of piracy were too tempting to resist. The story of the most famous Somali pirate, Mohamed Abdi Hassan, known as Afweyne ("Big Mouth"), who is based in Hobyo, is a story of entrepreneurial zeal. Afweyne was a former civil servant who took to piracy because it was a good business opportunity. In the absence of other ways of making serious money or a state willing or able to stop him, he began recruiting sponsors. One potential candidate, interviewed by Stig Jarle Hansen, ruefully remarked that "Afweyne started up in 2003. He asked me to invest $2,000, as he was gathering money for his new business venture . . . I did not invest and I regret it so much today." Afweyne traveled to more established pirate towns up the coast in Puntland on recruiting trips, headhunting men with the best reputations, and he quickly became the driving personality and business brain that turned what had been an amateurish, small-scale phenomenon into a well-financed, well-equipped operation run by professionals. Afweyne also played an important role in the opening of a pirate stock exchange. Based in the nearby town of Harardhere, it allows investors to buy shares in various pirate outfits and in particular attacks on high-value targets.

The pirates have stamped their authority and their image on Hobyo, transforming it from a port city with a complex and rich heritage into a caricature of lawless greed. It is not a good advertisement for how places cope in the absence of government. This is a point worth making because, according to one study by Professor Peter Leeson, a global authority on the economic impact of piracy, across a range of development indicators Somalia under anarchy has done better than its neighbors under bad government. Leeson also speaks enthusiastically about "the substantial increase in personal freedoms and civil liberties enjoyed by Somalis since the emergence of anarchy."

It's true that pirate towns of the past were often claimed to be relatively well-off and to allow a certain rough liberty. One of the reasons that Port Royal, the pirate haven in Jamaica that flourished in the seventeenth century, was protected by the island's governor was that it generated more wealth than could be expected from plantations. It seems that it was only when the pirates of Port Royal starting targeting local trading vessels that attitudes began to turn hostile. Port Royal was small compared to Canton in China. The pirate fleet based in Canton in the early nineteenth century comprised about four hundred junks and sixty thousand men. It *was* the economy. In the past there appears to have been some ill-defined tipping point at which such "feral" activity becomes dominant and everyone in some way becomes dependent upon its success. Yet today riches rarely stay where they are made. Professor Leeson's optimistic scenario of Somalia's ungoverned liberty comes unstuck in Hobyo. It's poor, decaying, and at the mercy of warring gangs of bandits. Although the sudden loss of piracy would probably hit some local people's pockets, Hobyo is merely a funnel for money extracted from foreign ship owners and passed on to Somali financiers who, according to Somali news reports, live in "beautiful buildings in Nairobi and Dubai" that are known as "pirates' buildings."

Being a feral city provides rich pickings for some, but it eviscerates the life of a place, and, moreover, it is a boom-and-bust existence. Hobyo was lost to Islamists in 2006, who were ousted six months later by Ethiopian troops supporting the regional government. The Islamists returned in 2009 and began to make life difficult for the pirates, although others accuse them of just demanding a large slice of the takings. More important, since 2011 Hobyo's pirates have also come up against better-defended commercial vessels and an international effort to destroy pirate bases, and as a result the number of ships taken hostage has fallen steeply. Hobyo's beleaguered mayor, Ali Duale Kahiye, feels that he might be getting his town back. Talking to journalists in 2012, he argued that "the decline of piracy is a much-needed boon for our region. They were the machines causing inflation, indecency and insecurity in the town. Life and culture is good without them." The regional government is talking up grand schemes to build a harbor and reintegrate Hobyo with the legitimate economy.

It would appear that the days of unfettered piracy are coming to an end. Afweyne recently announced his retirement from the business. The pirate towns of Somalia are more likely to be remembered as alarming exceptions than as harbingers of a new world disorder. Feral places are collapsing places, warred over, exploited, and weak. Hobyo's pirates could have been brought to heel many years ago if other governments had considered the town's fate important enough. Although the determination of individual pirates and the extraordinary rewards of their trade go a long way to explain Hobyo's pirate years, places only stay feral because they have been cast adrift by an indifferent world.

The absence of normality in spaces of exception makes them vulnerable: they often appear to be outlandish and inherently outside the spirit, if not the practice, of the law. Yet many established states and institutions started out this way. The struggle to make the transition from being seen as exceptional to acceptance and normalcy connects many of the places we encounter in the next section.

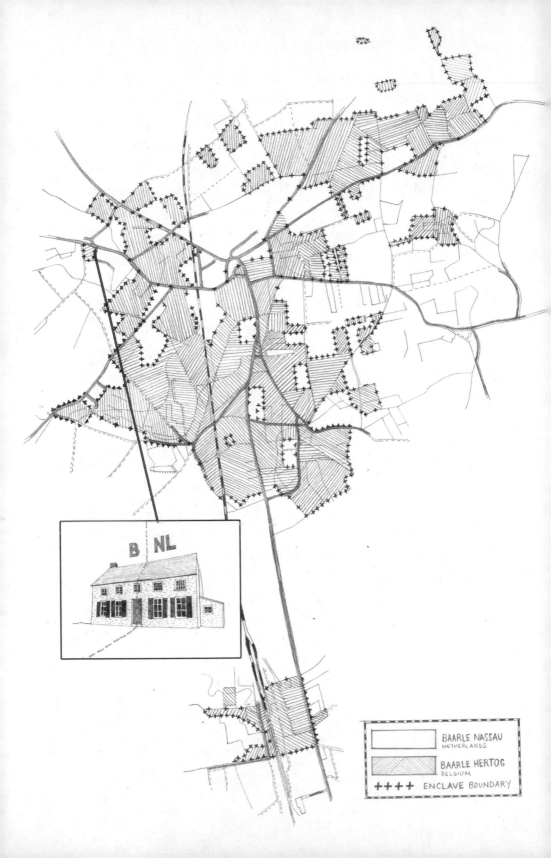

B | NL

BAARLE NASSAU
NETHERLANDS

BAARLE HERTOG
BELGIUM

++++ ENCLAVE BOUNDARY

ENCLAVES AND BREAKAWAY NATIONS

Baarle-Nassau and Baarle-Hertog

51° 26′ 20″ N, 4° 55′ 56″ E

I don't have an easy relationship with borders. They frighten and unnerve me. Searched, prodded, delayed; again and again, for the temerity of crossing a few feet of land, they are bureaucratic fault lines, imperious and unfriendly. It's not surprising that so many look forward to a world without borders. Their existence is routinely critiqued by academic geographers who cast them as hostile acts of exclusion. And yet where, in a borderless world, could we escape to? Where would it be worth going?

The possibility of new sovereign places depends on the creation of new borders but also provokes questions about their meaning and consequences, both in the embryonic ethnic nations of the world and in much smaller and quixotic stabs at autonomy. But the problems and pleasures of borders can also be witnessed in the fate of national enclaves, some of which suffer from a surfeit of bordering while others appear to relish it. For borders are far more than lines of exclusion— their profusion reflects the varied nature of people's political and cultural choices. The paradox of borders is that they close

down free movement yet suggest a world of choices and possibility.

For all their faults, there is something exciting about the
way borders snake over the land, about their power to impose
ideas and history upon the dumb earth. Perhaps this is what
Frank Jacobs, whose "Borderlines" column in the *New York
Times* excavated a rich seam of cartographic curiosities, was
getting at when he spoke of his sense of loss at the disappearance of borders. It's a syndrome Jacobs calls "Phantom Border
Sadness," which he defines as "a slight pang of grief caused
by the conviction that a world with one less border is also a
bit less special." In an era in which we are constantly urged
to pull down the barriers that separate us, it is a dangerous
thought. Yet Jacobs's eccentric nostalgia feels oddly humane,
for it acknowledges some things that are rarely acknowledged:
that people like creating borders, that they are not just frustrated by them but also thrilled and inspired.

Baarle-Nassau and Baarle-Hertog are two villages that sit
within and alongside each other. There are 22 bits of Belgium
(Baarle-Hertog, population 2,306) scattered in odd profusion
inside the Netherlands, within and around the Dutch town of
Baarle-Nassau (population 6,668), and 8 parts of Baarle-Nassau that sit inside these Belgic fragments. Some are block-
shaped but others are spindly creatures that sport long and
crooked tentacles. The largest enclave in Baarle is 1.54 square
kilometers, and the smallest, an empty field, is 2,632 square
meters. Of the world's 260 enclaves, about 12 percent exist in
and around Baarle.

The profusion of these borders means that when visitors
walk around Baarle they can never be quite sure which coun-

try they are in. It's hard not to conclude that these two places share the same space. This was certainly my experience, on the wet September day I went to Baarle, even though I was being guided by the municipalities' fractal town map. While some of the borders are marked, by white painted crosses on the pavement, there are just too many to make it practical to create signage for them all. On one 160-meter stretch of Kapelstraat, where visitors leave one large Belgian enclave only to pass through the borders of two nearby rectangular Belgium slivers, I was able to walk in a straight line across five national borders in under a minute.

Baarle is a friendly, workaday sort of place, and the residents take an unostentatious pride in being the world's only significant remnant of medieval border problems. Almost everywhere else borders have been straightened and rationalized, anomalies have been dealt with and forgotten about. The origins of the Baarle mix-up come from a time when enclaves were regularly thrown up across Europe as a result of the complexity and fluidity of local aristocratic domains and territorial claims. One exasperated eighteenth-century description of the French territory of Lorraine describes it as "mixed, crossed, and filled with foreign territories and enclaves belonging with full sovereignty to the princes and states of Germany." As a result, premodern maps were scribbled with borders. What this quote also shows is that, by the eighteenth century, enclaves were being seen as a problem. France applied delimitation treaties and old-fashioned conquest to eradicate many of them. The rational world of the Enlightenment tried to sponge away the dark and unmanageable world of enclaves. It left us with the view we have of them today, as oddities that somehow both demonstrate and escape the logic of the nation-state.

That Baarle survived is little more than good fortune. After an exhaustive examination of every last marriage, divorce, covenant, and claim behind the Baarle enclaves, the preeminent expert on the topic, Professor Brendan Whyte, simply shrugged his shoulders. Baarle's "integration into Napoleonic France," he tells us, "could easily have resulted in the rationalization of the enclaves at Baarle, as had been the case for most enclaves along France's northern and eastern borders," but "for some reason," this didn't happen.

Baarle is an exception, never important or irksome enough to get to the top of anyone's to-do list. Having survived, it now provides us with a living laboratory of medieval micro-borders. In 1959 the Belgian cattle dealer Sooy Van den Eynde challenged the Dutch town of Baarle-Nassau over its title to what he claimed was historically Belgian land. His case went to the International Court at The Hague, which ruled in his favor. As a result, a new Belgian enclave of about twelve hectares was created. In 1995 a border commission, after fifteen years of work, proclaimed that the Baarle borders were now known and fixed. However, the logic of fragmentation cannot be so easily tamed. In Baarle the incentive to find novel ways of complicating an already complex situation has taken on new life, which may feed an endless desire to define and scale down the borderline. How many centimeters thick is it? What can it pass through? In Baarle the custom has been that a property belongs to the country in which its street door is located. But what if the border runs through the door? In that case, the unsatisfactory outcome has been that the two parts of the building belong in different nations. This potentially tricky situation hasn't tended to last long in Baarle because of another local custom, namely, that tax is paid to the country where one's front door is. Not surprisingly, residents living on Baarle's numerous borderlines moved their front doors, shifting them a few feet to the cheaper country.

Door-shunting has died out over recent years but its memory is kept alive, along with a variety of border markers, including the helpful local habit of having the national flags on house number plates. The two village councils have alighted on Baarle's plethora of enclaves as its best chance of drawing in tourists. They are now working together to get the two towns listed by UNESCO as a World Heritage Site. It's a worthy bid, but I'm not sure Baarle will ever attract many visitors, for, apart from the quirkiness of the borders, there isn't a lot to see. "You are a tourist?" exclaimed one shopkeeper with mild shock when I told her why I had come. Yet there is no mistaking the pleasure that the people in Baarle take from their borders; they don't need tourists to remind them that they live in a cartographic legend.

Baarle is something of a best-case scenario for other, less happy, border squabble spots. It shows us how people can use borders to build a positive sense of self without making other people's lives a misery. Interviewed by the BBC a few years ago, the former mayor of Baarle-Nassau, Jan Hendrikx, clearly took pride in the fact that "our citizens mingle with the citizens of Baarle-Hertog, our Belgian neighbors, but not in a regular way." His equivalent in Baarle-Hertog, Jan Van Leuven, elaborated on the point. "My head is, I think, a little Dutch, but my heart is Flemish," he said, before concluding, "In general Dutch people are more rational. They think. They look to the north. We Flemish think also, but we are more emotional. We speak the same language but the words have a different significance."

It's easy to dismiss these kinds of generalizations, especially when Baarle's borders have been made harmless by both Belgium and the Netherlands being in the European Union. When Frank Jacobs had the gall to write in praise of borders for the *New York Times* he received a severe scolding from some of his readers. Didn't he know that "borders are for small minds which exploit fear and ignorance, and attempt to circumscribe a human species which knows no limitations?" Didn't he understand, wrote a self-styled "professional geographer," that professional geographers have discovered that borders are nothing more than "old-fashioned colonialism"?

Thankfully not. Of course, enclaves are a serious problem in some parts of the world, as we shall see next in Chitmahals and observed earlier in Ağdam (see page 108). In fact, in a curious return to medievalism, they started emerging again in the 1990s as the USSR fell apart and some twenty new enclaves were created. More could follow. The likelihood of a world without borders is not high, and when we think of that world—a utopia of sameness where there is no possibility of escape—we might begin to wonder if it is an attractive destination. Baarle illustrates how creating places and creating borders are interwoven. It also provides an example of the pleasure of borders, hints at their playful and absurd side, and shows us how, in a topophobic world, the fascination of the border can be rediscovered and humanized.

Chitmahals

26° 16' N, 89° 06' E

Chitmahals are the "paper palaces" of the India–Bangladesh border. A composite name, it combines the English word for a monetary slip—a chit—with the Hindi word for palace, "mahal." Nearly two hundred of these enclaves are chaotically nestled in and around the border zone. They range in size from Balapara Khagrabari in Bangladesh, at 25.95 square kilometers, to Upan Chowki Bhaini, a counter-enclave (an enclave inside an enclave) and, at 53 square meters, one of the smallest enclaves in the world.

Local folklore has it that the chitmahals are the result of aristocratic play. It is said that the chitmahals, which are often referred to by their Indian regional name of Cooch Behar, are spots of land won or lost in a chess game between the Maharajah of Cooch Behar and the Nawab of Rangpur in the early eighteenth century. A rival story pushes the date of their creation into the mid-twentieth century. In 1947, a drunk British cartographer is said to have knocked over his inkpot while finishing this tricky section of India's provincial borders. All the specks and blotches had dried by morning and ended up as the world's most pockmarked geopolitical landscape. While there's no truth to the second legend, the first captures the key points. The chitmahals are indeed remnants of princely deal-making and favors dating from the early eighteenth century. But for most of their history they were far less sealed off than they became in the second half of the twentieth century.

The creation of India, then Pakistan, and then Bangladesh turned the chitmahals from local anomalies into an international problem.

Unlike Baarle-Nassau and Baarle-Hertog (see page 177), the 51,300 people who live in this geopolitical jungle don't revel in their uniqueness. The chitmahals show what happens when borders subvert rather than foster people's sense of place. The residents aren't proud of their baroque borders, their numerous counter-enclaves (there are twenty-eight of them), or even having the world's only counter-counter-enclave, which is called Dahala Khagrabari and consists of seven thousand square meters of India inside a Bangladeshi village, which sits inside an Indian enclave in Bangladesh. No one is interested in the curiosity value of all this because life inside the enclaves is grim. For the chitmahals are not places of tourism or creativity but of abandonment and imprisonment.

Most of the enclaves have been left to fend for themselves, which means that the residents have had to build their own schools and bridges. Each enclave has had to come up with its own system of justice and enforcement and its own way of settling disputes about land ownership and just about everything else. It is anarchy, a self-governing society without government, and it means that children don't get educated, roads and bridges are ramshackle, justice is meted out by the strong against the weak, and disputes between residents turn into vendettas. As a result, the villages look with envy at their neighbors, who live in places protected and ordered by states. As one thirty-eight-year-old who sits on one of the local councils that run many of the enclaves explains,

"If there is a flood in Bangladesh, or any other country, relief will come. If we are surrounded by water we will have to die here. When it floods people are like prisoners surrounded by water."

The problems of the chitmahals are caused by too many borders preventing the state from functioning adequately. It turns out that when you cut off people, especially poor and vulnerable people, from the state, you inhibit rather than enable autonomy. The proof can be found within these blotchy borderlands. Their plight shook loose the last remnants of my anarchism, or at least the vague antistatism I quietly nurtured well into middle age. My cozy faith in community autonomy, shaken by the bleakness of Hobyo (see page 168), has been snuffed out by the chitmahals. Now when I read the nineteenth-century anarchist Mikhail Bakunin, I wince. "If there is a State, then there is domination, and in turn, there is slavery," he claimed in 1873. And if there isn't a state? It usually turns out to be worse.

But it's also true that the residents of the chitmahals have not been treated well by the states of India or Bangladesh. Their life has been made hard, cruelly so. Indeed, it could be said that their key problem is one manufactured by state bureaucracy. In order to leave these tiny enclaves, the inhabitants have to obtain a visa to travel through the foreign territory that surrounds them. But in order to obtain a visa they have to leave their enclave, since visas can only be obtained in cities many miles away. The consequences of this system have frequently been tragic. In 2010 a young woman called Jaimona Biwi, a resident of Mashaldanga village, a Bangladeshi enclave inside India, started to go into labor. She traveled to the hospital nearest to her chitmahal, but because she was not an Indian citizen the doctors refused to see her. "I was in unbearable pain but lay on the hospital floor for two days," she explained to Indian journalists. "I bled profusely and delivered a stillborn child." When she became pregnant again she relied on a local trick, hiring an Indian husband. But she is bitter: "He has taken 2,000 rupees from our family and has given his name in the hospital register as my husband." Biwi's words hint at the stigma that comes with living as political untouchables. One of the few Indians who has moved into Mashaldanga says, "In our area, no resident of an Indian village will marry a man or a woman who is from a chitmahal. Only if a family is very poor will they marry their daughter to someone from an enclave in exchange for a bride price."

It is hoped that recent moves by both countries to cede territory will soon allow the chitmahal problem to be resolved. A land exchange deal signed in 2011 should see most of the enclaves merging with the nation that surrounds them. When this finally comes to pass the number of enclaves in the world will plummet by 70 percent. They won't be missed by many of the people who live in them, many of whom want to change nationality. This is partly a consequence of the fact that the chitmahals are largely populated by families who fled their own country. A seventy-year-old Indian Muslim in the Kajaldighi enclave, inside Bangladesh, interviewed by political geographer Reece Jones, recalled that he and his family came there in 1969 because "there was violent oppression of Muslims at that time in India. There was lots of fighting, killing, and extortion. We were threatened by local landlords. In order to preserve our respect, honor, and lives we fled. Everyone came." He went on to say, "At that time everyone in the enclaves was Hindu. When we came here, they left." This old man would be delighted if his enclave was absorbed into Bangladesh. But others are less happy about this prospect. The Indian Enclave Refugees' Association has been formed to lobby for the right to "return" to India. This group asserts that "refugees" from the Indian enclaves in Bangladesh trying to return to India are treated as foreigners in their home country. They are spurned by India and denied the right to settle. One of the group's spokesmen, Deb Singha, complains, "Dwellers of Bangladeshi enclaves in India will now be called 'Indians' and we, despite being Indians, don't have that right."

The endgame for the chitmahals may be a messy business. And at each turn the states involved, rather than being helpful, have either turned a cold eye on the plight of these people or gone out of their way to make their lives a misery. The residents of the chitmahals experience a common predicament but in a uniquely stark form. These places are abandoned by the state but it's the state that hounds them. Their paper palaces are islands of freedom from the state but also prisons within it. The paradox remains that government is essential to our liberty but it also pushes us around and demands our subjugation. All of us live with this dilemma in some form; it is an expected facet of modern life that people both need the state yet resent it. Some manage to escape this uneasy relationship with the government by striking out on their own. But setting up a breakaway nation poses its own challenges.

Sealand

51° 53′ 40″ N, 1° 28′ 57″ E

Wouldn't it be fun to have your own island nation with your own laws? No matter how small, on an island your independence is tangible. Your borders are real and beyond doubt. With a bit of luck your kingdom will be unclaimed and there waiting for you, requiring you only to hop onshore, unfurl your flag, and claim sovereignty. Just like Sealand.

The Principality of Sealand is an "independent state" that was established in 1967 by a retired British army major, "Paddy" Roy Bates, on an abandoned World War II gun platform. It is located on Rough Sands, off the Essex coast, and has an inhabitable area of 550 square meters. The platform was built in international waters and abandoned after the war, so when, on September 2, 1967, Bates and his family hauled themselves up the side of the old sea fort, they felt entitled to claim sovereignty. Bates bestowed upon himself and his wife the titles of Prince and Princess, and "the royal family" set about turning Sealand into both a home and a kingdom.

Sealand is the best example of a sub-breed of the world's numerous self-started nations, or micro-nations, called platform states. These maritime nations are often short-lived. Ernest Hemingway's brother, Leicester Hemingway, was briefly president of the bamboo-platform republic of New Atlantis, off the west coast of Jamaica, anchored down by a railway axle and a Ford engine block. When this structure was destroyed in a hurricane in 1964, Hemingway created another such fiefdom, Tierra del Mar, an 840-square-meter platform built on a sandbar near the Bahamas. However, the US State Department persuaded him not to declare sovereignty; it was worried that Tierra del Mar could serve as a springboard for the annexation of nearby islands by unfriendly powers.

Compared to such fragile endeavors, Sealand has a long and rich history. In 1968 the Royal Navy sent two gunboats to evict the Bates family. Roy Bates fired warning shots at them and was arrested. However, he managed to argue in court that British law did not apply to Sealand. In its judgment of November 25, 1968, the court declared that Sealand was outside of British national territory and hence jurisdiction. Flush with success, the Sealanders began to issue gold and silver coins, the Sealand dollar, and postage stamps. The sale of these collectors' curiosities helped to finance the new kingdom. Passports were also issued, but whether these too were offered for sale is disputed. The next dramatic event in Sealand history came in 1978 when a group of Dutch and German businessmen visited with a commercial proposition. While they were there, they took the fortress at gunpoint, holding Roy Bates's son and heir, Prince Michael, prisoner for three days. He was freed only after his father organized a counterattack. The *Sealand News,* the island's online newspaper, reported Bates as saying, "We phoned a friend who owned a helicopter." The story went on: "He'd performed stunts in James Bond films, but this was the first time he'd done it for real. 'I descended from the helicopter to the platform with a gun, fired a shot and said: "Everyone put their hands up!" and that was that.'"

There was plenty of gunfire but no injuries, and in true Bond style, the invaders were swiftly overpowered and taken as "prisoners of war." When the German government asked Britain to intervene, it was told the fortress was beyond British jurisdiction. The prisoners were later released, after the arrival of a German diplomat. The first to be freed, though, were the Dutch citizens. The one German invader was held a little longer because, it was reasoned, he had accepted a Sealand passport and therefore was guilty of treason.

Those were Sealand's glory days. But most games, however serious, lose their excitement in the end, and the players drift off. Despite its military triumph, in the 1980s Sealand began to sink into disrepair and disuse. Its independence was also called into question when, in 1987, the UK extended its territorial waters from three to twelve miles. Sealand retaliated with its own twelve-mile claim, declaring the annexation of Harwich and Felixstowe on the English coast.

Britain has made no attempt to take Sealand, and the British government still treats it as a de facto independent state. Prince Roy paid no British National Insurance during his time as a resident of Sealand. Moreover, the obvious infringements of UK law on Sealand continued to go unpunished. For example, when Roy Bates fired "warning shots" at a ship that he felt was straying too near his kingdom in 1990, the ship's crew's complaints were not pursued. In its official history of the new state, the Bates clan claims that this was "a clear indication that Britain's Home Office still considers Sealand to be outside their zone of control."

Unfortunately, the buccaneering image of Sealand drifted into murkier waters in the 1990s. In 1997, after the killer of fashion designer Gianni Versace committed suicide on a Miami houseboat, the police discovered that the man who owned the houseboat had a Sealand passport. In the spring of 2000, Spanish police arrested a Madrid-based gang tied to international drug trafficking, money laundering, and selling forged documents. Found among its cache were thousands of Sealand passports. Questioned by Interpol, Roy Bates was adamant that the passports were not for sale. "Sealand has all been a game, an adventure, and it is very unfortunate to see it take this turn," he told one reporter. The source of these forgeries was later traced to the same German man who had tried to take Sealand by force nearly twenty years earlier. Dubbing himself Sealand's Minister of Finance, he had created a fake Sealand Business Foundation and boasted that he had sold more than 150,000 fake Sealand passports. The Sealand national website claimed that "many of the forged passports were sold to people leaving Hong Kong at the time of Chinese reoccupation, for USD 1,000 each."

The lease of Sealand to an Internet company, HavenCo, in 1999, compounded its tarnished image. Silicon Valley investors pumped in hundreds of thousands of dollars in seed money to turn Sealand into an off-limits Internet server and "fat-pipe data haven." Child pornography, spamming, and malicious hacking were prohibited, but with no restrictions on copyright or intellectual property for data hosted on its servers, Sealand looked set to become a center for file sharing and other infringement activities. The injection of cash was used by the Bates family to secure the kingdom. A desalinization plant and new generators were installed, satellite connections were established, microwave links to Britain were rigged up, and miles of cables were laid. With the help of the extra cash Prince Michael also established a Sealand army and navy, with machine guns and high-speed dinghies. Despite all this, HavenCo's customers soon began to thin out. What seems to have scared away most of them was the fact that all of Sealand's Internet traffic went through Britain, which claimed that the platform was within its territorial waters. In 2008 HavenCo went bust, and the media was fed a story by Michael Bates that Sealand was up for sale, valued at 750 million euros. This appears to have been a ploy to drum up interest in a new leasing arrangement. With no new funds, Sealand became, in the late 2000s, once again uninhabited, and Prince Michael lives out his days in Leigh-on-Sea. This is also where his father spent his last days. Roy Bates died on October 9, 2012, at the age of ninety-one.

Talking to the *Sealand News,* Joan Bates offered a romantic view of life on Sealand: "It's been a fairytale. What greater compliment can a man pay to a woman than to make her Princess of her own Principality? I love being able to call myself Princess. When we travel abroad on our Sealand passports we are always greeted with a lot of fuss and treated like royalty."

It's a fantasy that is immediately comprehensible. For those of us who are bored and frustrated by the endless rules and impersonal bureaucracies of conventional nations, the idea of creating one's own island kingdom has tremendous appeal. Indeed, the most telling recent stories about Sealand concern the way it has been taken up as an icon by a new generation of eco-libertarian planners. The Seasteading Institute, founded in 2008 in California, looks to Sealand as a pioneer. The aim of the institute is plainly stated as "creating sustainable sea platforms where people can choose to live if they're unsatisfied with life on solid ground." Man-made sea kingdoms have a growing appeal, and something of that appeal is being captured in new islands built in the Maldives and Dubai, as well as in a new generation of cruise ships that are morphing into permanent sea-roving settlements. But the root of Sealand's appeal takes us into terrain that these ventures still only hint at. It is the territory of independence, of sovereignty, of really having one's own place. It may be a foolish fantasy, but it's an important and very human one.

United Kingdom of Lunda Tchokwe

12° 39' S, 20° 27' E

Although it is accompanied by a lot of polemic about fragmentation and crisis, the birth of new nations is an unstoppable and unending process. One of the reasons it has become a difficult and sensitive topic is that ethnicity is still central to most acts of nation-building. The widespread tendency to treat such ethnic claims as throwbacks consigns many embryonic nations to the geopolitical shadows. This is where the United Kingdom of Lunda Tchokwe languishes, ignored by the outside world yet ceaseless in its struggle.

Lunda Tchokwe covers the eastern half of Angola. If the kingdom is known to people outside Angola at all, it is for one thing: being the home to some the world's largest diamond mines. An area a little smaller than Spain, with a population of only four and a half million, it is one of dozens of unrecognized countries in Africa and one of twenty-one that belong to the Federation of the Free States of Africa. It is a key member of the inner sanctum of that group, being part of the eleven proto-states that have formed an Economic and Defense Alliance that claims various portions of a huge area of southwestern Africa. Although the succession of South Sudan has bolstered the confidence of the federation, its members need one another because no one else is in the slightest bit interested. It's an absence of concern that reflects a suspicion of ethnic secessionism, because if the Lunda Tchokwe get their own country, then why shouldn't every other ethnic group in Africa? If such a thing came to pass, it would make the disintegration of the USSR look unspectacular: the continent could easily be covered with thousands of nations.

Yet these nascent nations are taken very seriously by the governments whose territory they claim. Angola has criminalized the activities of the Lunda Tchokwe separatists and forced many activists into exile. Nearly 40 members of a group that goes under the cumbersome name of the Commission of the Legal Sociological Manifesto of the Lunda Tchokwe Protectorate were accused of "destabilizing national order" and arrested between April 2009 and October 2010, along with 270 suspected supporters. Many were thrown into prison. One member, Don Muatxihina Chamumbala, subsequently died, and is today hailed as "the first martyr to die in defense of the natural rights of the people of the Lunda Tchokwe." The commission accuses the unelected Angolan regime of numerous abuses of human rights and of leaving the region to rot. The movement's website claims that the "population of the Lunda Tchokwe are well aware of the state of total abandonment that this diamond rich area is in."

The story of Lunda Tchokwe illustrates a modern paradox: the remorseless power of ethnic nationalism in a world where it is increasingly believed that national identity should have nothing to do with ethnic identity. The nation-state may have grown out of ties of language and culture, but its contemporary, bureaucratic form is supposedly postethnic, or at least able to accommodate ethnic diversity. To be British or American is a matter of holding the right passport rather than having the right heritage. The claims of the world's unrecognized states threaten to throw this cosmopolitan dynamic into reverse, and the rhetoric of the Federation of the Free States of Africa is crystal clear on the failure of what it regards as pseudo-countries like "Angola, Nigeria, Senegal, and Kenya." On its official website, the organization's secretary general, Mangovo Ngoyo, explains that these nations "will always [have] problems, because they do not constitute a country such as England" but rather "several countries with their specific culture, national identity, own separate language, own architecture, own history." Ngoyo's unfortunate choice of England as his example proves the awkwardness of his thesis. England hasn't been a country for over three hundred years, having combined with Scotland in 1707 and developed into a modern, multiethnic United Kingdom, which accommodates wide variations of culture and heritage. That doesn't mean that ethnic nationalism, in England or anywhere else, has gone away, but rather that it survives in a fraught relationship with other forms of association. The demands of the unrecognized states throw this unresolved and difficult relationship into relief. Indeed, the Federation of the Free States of Africa gives much play to the statement by the British prime minister, David Cameron, that "state multiculturalism has failed." Ngoyo adds, "Of course multiculturalism is bound to fail. A Nation can only be a Nation if all are singing from the same 'Chorus Book,' if not then there is no harmony."

Ngoyo also asks, "How are we expected to keep 'Colonial Marked Border States' in harmony?" It is this last point that brings us to the nub of the issue for nationalists in Lunda

Tchokwe. They don't just see "multicultural" Angola as a failure. For Lunda Tchokwe activists, Angola is a colonial power,
its colonialism an extension of European colonialism. They
see their lands stripped of natural resources and the profits
funneled away. Once they went to Portugal, the area's former
colonial master, and now they head to the boomtowns on Angola's east coast. The Angolan government makes much use
of antiseparatist counterarguments once deployed by the Portuguese: that indigenous resistance is a symptom of tribalism,
and that the peoples of Angola need to be saved from factional
warfare. They are able to give this old line a modern spin by
claiming that Angola is a multicultural, and hence modern,
liberal state, and that the nationalists are xenophobes. It is a
galling accusation for activists who are routinely thrown into
jail because of their ethnic affiliation.

In fact, Lunda Tchokwe has a more complicated relationship to multiculturalism than some of the rhetoric that comes
out of the Federation of the Free States of Africa implies. The
ambition of a "United Kingdom" points to the fact that the
Lunda and Tchokwe were once two separate groups and to
the long history that has brought them together. The Lunda
Kingdom had spread over 150,000 square kilometers by 1680
and kept on growing, and by the end of the nineteenth century it had doubled in size. In the course of this expansion the
kingdom grew into a federation of restless clans. One of these
tribute-paying clans, the Tchokwe, rebelled against Lunda
rule and by the end of the nineteenth century had effectively destroyed the old kingdom. The contemporary nationalist movement places much store in Portugal's recognition of
Lunda Tchokwe during this period and in the signing of various protectorate treaties between Portugal and local kings.

However, it's not a particularly stable legal lineage. It was only because of the ongoing dismantling of the Lunda Kingdom that the Portuguese found it so easy to extend their empire eastward. Moreover, this was always remote territory. Portugal's real relationship was with its seaboard colony, which was founded in 1575. The far-eastern interior was beyond its control and interest. Some authorities claim that it was not until the 1930s that the Portuguese even came into contact with the Tchokwe. Certainly it was only from that decade that Portugal successfully absorbed the area into the established colony of Angola, which became independent in 1975.

This potted history tells us that, unlike Angola, Lunda Tchokwe is able to lay claim to a rich, complex, and independent African history. It also shows that Lunda Tchokwe's relationship to Angola is both recent and shallow, and that far from being culturally homogenous, the Lunda Tchokwe are a diverse group. The various peoples of Lunda Tchokwe have fashioned a common sense of place and of group allegiance. This identity is a very recent creation but it still matters, for by attaching people to a particular part of the world, it anchors and sustains a shared vision of the past and the future.

At the moment, the chances of any story about Lunda Tchokwe making it onto the global news agenda are slim. Speaking out against the regime is too dangerous, and Lunda Tchokwe has no military or insurgent forces. Angola is not going to let itself be torn in half, and there isn't a state in this part of Africa that doesn't support its crackdown on secessionism, for this is a force that threatens every one of them. We can safely conclude that the United Kingdom of Lunda Tchokwe will not appear in our atlases any day soon. But new nations will continue to be born, especially in those parts of the world where ethnic and territorial claims have been steamrollered by history. In this context, the need to claim and defend one's own nation is constantly being reimagined and rediscovered. The struggle to create such new places is hugely difficult, but so too is the struggle to keep composite colonial creations like Angola together.

Gagauzia

45° 05′ N, 28° 38′ E

The story of Gagauzia tells us about the remorseless power of nationalism to keep dividing and subdividing nations into smaller units. Gagauzia is in the south of Moldova, a small landlocked country of three and a half million people that is wedged between Ukraine and Romania. Moldova broke away from the USSR and became independent in 1991, but it is a patchwork of nationalities that shows every sign of becoming unstitched.

The map of Gagauzia is a ragged thing. This would-be state is spread across four unevenly sized enclaves within Moldova. In all, it covers an area of 707 square miles, about half the size of Rhode Island, and has a population of 161,000. It's never going to be a giant among nations, but a people's desire for freedom is not proportional to their number or the size of their territory. The force that causes nations to fly apart has a centripetal energy: it is a creative and unpredictable dynamic that gives birth to new demands for independence at the very moment it answers the demands of others. It's a mistake to patronize places like Gagauzia or cast them as the off-shoots of chaotic regions, for the fragmentary logic at work here is at work elsewhere.

National independence is not a one-off event, a book that once opened can simply be closed. It may be comforting to think that, for example, once Scotland is independent, then a long tale will have reached its happy end. But nationalism spills out, catches on, transmutes other place-based identities into nation-building projects. If Scotland is independent, then why not Shetland? If Moldova is independent, then why not Gagauzia? Nation-making is a process that does not simply fulfill needs; it also creates them.

One of the few people who have studied Gagauzia is the Turkish anthropologist Hülya Demirdirek. Even she is a little mystified by the self-invention of the Gagauz people into a national entity called Gagauzia, a word and an idea that barely anyone had heard of twenty years ago, because until the USSR broke up, no such place existed. At a conference on "post-communist anthropology," Demirdirek conceded that "it is difficult to answer the question of who the Gagauz think they are." One conventional answer is to say that the Gagauz are Eastern Orthodox Christians who trace their ancestry to Bulgaria and who speak Gagauz, a language that is similar to modern Turkish. They are a distinctive mix of the Christian and the Turkic, with some Gagauz claiming that they were the founding people of Bulgaria, descendants of the Bolgars who conquered that country in the ninth century. However, a more pertinent aspect of their complex heritage is that the Gagauz are one of the most culturally Russianized groups in Moldova, with many preferring to speak Russian rather than Gagauz. It was an unfortunate association for the Gagauz, be-

cause Moldovan nationalism is defined around an antipathy
to the country's former Soviet masters. As Moldova's independence grew nearer, the Gagauz found themselves increasingly
portrayed as a foreign element, a people apart whose real loyalty was to Mother Russia.

It was in this hostile atmosphere, in 1988, that a social
movement called the Gagauz People was founded and began to demand independence. A Moldovan Parliament report from 1990 alarmed the movement further by naming the
Gagauz not as a national minority but merely as an "ethnic"
minority. It was a choice of words that was widely interpreted
as a calculated insult. Some members of the Moldovan Popular Front went so far as to demand that the Gagauz, like the
Russians, should go "back home." It was around this time that
"Gagauzia" was invented.

The desire to reinvent a place as a nation does not necessarily emerge from long-repressed aspirations but can arise
suddenly, especially among vulnerable populations whose
identity was once absorbed by vast, multinational entities
like the Soviet Union and who now nurse a sense of being
discriminated against and overlooked. With that perceived
slight, a number of useful myths were born. It was said that
Gagauzia had long been repressed, that the Gagauz had long
yearned for freedom. Some even argued that they were not
of foreign extraction at all but had been in this part of the
world longer than the Moldovans. Little of this was true, and
Gagauzia was far from ancient. Aside from a five-day independent state, declared in 1906 and limited to the capital (the
Republic of Kormat), the Gagauz have never thought of themselves as needing their own nation.

Yet this lack of historical depth only seems to have piqued

their political aspirations. In 1990 the unofficial Gagauz flag, a dramatic red wolf's head on a white circle, appeared on state buildings, and in August of that year independence was declared. Presidential elections were held and a government was installed in Kormat. Over the next four years Gagauzia claimed to be independent, although no other state deigned to recognize its existence. By late 1994 Moldova was willing to concede "self-determination" to "the people of Gagauzia," and a referendum was held that resulted in the present hotchpotch of enclaves, with thirty settlements voting in favor of being in the newly created "national autonomous territorial unit."

These concessions don't amount to much, since the only real gain is a promise that the Gagauz can decide to go it alone if there is a "change of status of the Republic of Moldova." The reason this matters is that if it wasn't for the national minorities in their midst, Moldova would probably opt to merge with Romania, with which it shares both history and language. The unification movement is one of the most powerful forces in Moldovan politics, and would turn the Gagauz from a small but vocal minority in a plural state into an "ethnic" irrelevance in a pan-Romanian nation. The Gagauz, along with the even more fiercely independent Transnistrians, who live on Moldova's eastern flank, have been promised that if that happens, they can leave.

However, the last twenty years have driven home to the Gagauz that being a "national autonomous unit" delivers very little. Gagauzia remains one of the poorest areas of Moldova, which in turn is often claimed to be the poorest country in Europe. The mood for compromise is waning while the momentum toward separation is being given new life by the development of an independent Gagauz media, led by Gagauzia

Radio Televizionu. In 2012 a Gagauz nationalist threw a Molotov cocktail at the motorcade of the visiting German chancellor, Angela Merkel. That same year Mihail Formuzal, the governor of Gagauzia, responded angrily to increased signs that Moldovan-Romanian unification was gaining popular support. Formuzal threatened to declare independence and boasted that, this time, his country would achieve international recognition.

There is an unnerving quality to the fragmentary logic of nationalism. Countries one has barely heard of break up into units that mean almost nothing. The logic of disintegration creates a geography of ignorance, in which the flowering of new identities and new nations outstrips our capacity to place or pronounce them. People outside the region throw up their hands: places like Gagauzia are consigned to a growing pile of ignored proto-states. Behind this reaction is an understandable fear: What if every nation started to be pulled apart and the political map resolved into legions of multiplying places? It may be convenient to imagine that we're above nationalist desires, that they are mistaken or somehow tragic. But such lofty dismissals are based on just as many myths and conceits as the fabricated pasts of Gagauzia. And they lack generosity. Many Gagauz want their own country because without it they will remain placeless and marginal. The fact that it is invented won't make it any less real.

FLOATING ISLANDS

Pumice and Trash Islands

Part of the attraction of floating places is their unplaceability: they promise escape from prosaic solidity and a freer relationship to the earth. Floating places have been on our minds since Aeolia, the floating island visited by Odysseus whose king was in charge of the four winds. When Gulliver visited Laputa, an airborne kingdom of distracted scientists, and Dr. Dolittle stepped onto the bobbing shores of hollow Sea-Star Island, they were joining a long tradition of geographical fantasy. It is an aspiration that has come into its own with computer games in which players skip between many islets. Further proof, if it were needed, that for earthbound creatures like ourselves, buoyant or untethered land is intrinsically enchanting.

So it is not surprising that the news that there are islands that drift on the sea was greeted with innocent joy. That two should apparently heave into view at the same time sounds doubly delightful, but what a strange contrast they make. One is a coagulation of plastic detritus known as the Pacific Trash Vortex. The other is a natural byproduct of volcanism known as a pumice raft. Neither is the Aeolia or Sea-Star Island of our dreams and even the name of the Trash Vortex sounds deeply sinister, but both remain oddly thrilling. Hence the urgent questions: Can we walk on them? Can they sustain life? The answer to the questions is "yes" for the pumice rafts, and "probably not" for the whirlpool of rubbish.

One of the largest pumice rafts ever recorded was found in 2012. Indeed, to call it a raft does not do it justice. This one, spotted by the New Zealand Air Force 620 miles off the Auckland coast, was spread over an area of 10,000 square miles, or "nearly the size of Belgium" as the New Zealand press described it. Naval lieutenant Tim Oscar said it was the "weirdest thing" he'd seen in his eighteen years at sea. In fact, smaller versions of such rafts are not that uncommon, nor are they confined to the Pacific, since they are caused by undersea volcanic eruptions. Oceanographers have charted their transoceanic voyages back some two hundred years.

Yet they seem to catch even seasoned ocean watchers and mariners by surprise. Making their way to Fiji on the yacht *Maiken,* the Swedish sailor Fredrik Fransson and his crew sailed into one in August 2006. Fransson's log describes the scene:

> We noticed brown, somewhat grainy streaks in the water. First we thought that it might be an old oil dumping. Some ship cleaning its tanks. But the streak became larger and more frequent after a while, and there were rocklike brownish things the size of a fist floating in the sea. And the waters were strangely green and "lagoon like" too. Eventually it became more and more clear to us that it had to be pumice from a volcanic eruption. And then we sailed into a vast, many miles wide, belt of densely packed pumice. We were going by motor due to lack of wind and within seconds Maiken slowed down from seven to one knot. We were so fascinated and busy

taking pictures that we ploughed a couple of hundred meters into this surreal floating stone field before we realized that we had to turn back.

Fransson goes on to explain how the pumice abraded the bottom of his boat: "Like we'd sailed through sandpaper." His photographs show miles of packed rock floating freely on the sea. Photographs taken by marine scientists indicate that the pumice can pile up quite deeply, sometimes extending several feet above the surface and forming an undulating landscape. It is at this point that the rafts can bear a human's weight, although so far it appears that this has been put to the test only near shorelines.

The rafts eventually drift toward land, clogging up the harbors they bump into. When they make contact with terra firma it becomes evident that they are far from barren. Clinging to the pumice fragments are a range of shellfish. One of the new scientific theories about how plants and animals spread around the world goes under the title of "rapid long-distance dispersal by pumice rafting." A study of the migration and inhabitants of the rafts spat up from the same undersea eruption that the *Maiken* sailed into found them to be teeming with life. Erik Klemetti, a professor of geosciences at Denison University, records what was found:

The pumice quickly became home to upwards of 80 different species of marine life over the course of its journey—in some cases, single pumice clasts [fragments of rock] were home to over 200 individuals of a single species of barnacle (this means that over 10 billion barnacles colonized the pumice raft). Some of these critters were permanent inhabitants (that is, they were attached) while others were mobile, so if the pumice landed on a beach, off onto the island a crab might scuttle. By a year and a half after the pumice raft was erupted, some clasts had ¾ of their surface covered. It could reach such an extreme that the biological hitchhikers would cause the pumice to sink or preferential float with one side facing up, creating microenvironments on a single pumice clast!

Klemetti concludes that "these volcanic events that have happened frequently in the recent geologic record all over the world may play an important role in how life colonized different parts of the world's oceans."

In recent years, stories about pumice rafts — especially the blog and photographs of Fredrik Fransson — have been working their way around the world in the form of "look at this!" email attachments. Everyone loves them. The Pacific Trash Vortex is a scarier but no less jaw-dropping prospect. Estimates of its size vary considerably. The truth probably lies between 270,000 square miles and 5,800,000 square miles. Since Belgium appears to have become an international standard for judging the size of large floating objects, we can also translate these figures as 22 times the size of Belgium to 592 times the size of Belgium — in other words, nearly twice the size of Australia. Also known as the Great Pacific Garbage Patch, the Trash Vortex does not exist as a single entity but is more like a soup or galaxy of garbage, most of which floats just underwater but often coagulates on the surface. Footballs, kayaks, and Lego blocks have all been spotted, along with the usual mass of plastic bottles and fishing nets. As Donovan Hohn revealed in *Moby-Duck,* it is also the resting place of a good many toy ducks.

All the things that are thrown off ships or swept up off the coasts of the Pacific and get caught in the circulation of the ocean waters end up in this graveyard of consumerism. As with the pumice rafts, one of the first accounts we have of this new landscape came from an adventurous yachtsman. In 1997 Charles Moore was on his way back to Los Angeles from Hawaii. He decided to take his yacht into a part of the ocean usually avoided by sailors because of its slow currents and lack of wind. To his astonishment he found himself sailing into a sea of gunk. "Every time I came on deck, there was trash floating by. How could we have fouled such a huge area? How could this go on for a week?" Moore, heir to a fortune from the oil industry, has since become a leading campaigner and researcher on trash vortices. Despite their size, and their human origins, they are little understood. Basic questions about how they move, how rubbish is sorted within them, and where that stuff then goes have yet to be answered. Oceanographer Curtis Ebbesmeyer has argued that they "move around like a big animal without a leash," and every so often they find a shore and cough up plastic all over the beach. Ebbesmeyer puts it in suitably grotesque language: "The garbage patch barfs, and you get a beach covered with this confetti of plastic."

Since all oceans have circulating currents, called gyres, and since rubbish is being picked up by such currents around the world, one would expect that trash vortices are forming in many oceans. In fact, the Pacific has two: an Eastern Garbage Patch has been discovered off Japan. But there is also the North Atlantic Garbage Patch, first identified in 1972. It drifts about one thousand miles in the course of a year. Rare photos of it have been taken by the research vessel *Sea*

Dragon, which since 2010 has been studying trash vortices all over the globe. They show a mass of floating debris on rough seas.

But we can't end the story there. A narrative arc that takes us from Homer's Aeolia to a gray gazpacho of plastic in the cold Atlantic may satisfy the modern urge for environmentalist misery, but it's too short a tale to do justice to our love affair with places that float, because even when they are made of rubbish they are still amazing. A very old desire has been whetted, one that, as we shall now see, is today being fulfilled and tested by manufactured floating islands.

Nipterk P-32 Spray Ice Island

One of the most remarkable of island-building technologies—the spray ice island—is likely to have only a brief moment of glory. Almost at the very moment it was perfected, it started to look like the debris of another era, a parable for our generation of the fate of the Arctic, which within decades will be completely ice-free in high summer.

In 1989 ExxonMobil built Nipterk P-32 in Canada's Beaufort Sea, at the frozen top of the North American continent, and while not the first of these islands, it was the biggest and most ambitious. The region holds vast oil reserves, but until recently, most of the year the Beaufort Sea has been entirely iced over, with only a small coastal channel leaking open between August and September. Making spray ice islands in such a subzero climate is simple, and starts by hosing water high into the air. The water freezes before it can hit the ground and builds up on the sea ice. In shallow waters, after many days of continuous spraying the sea ice is weighed down to the ocean floor, so it's not a true floating island for long. The hoses remain on until a roundish island is formed well above water level.

Ice islands take a variety of forms, but the best-known ones are made by nature. When ice shelves shoved forward by glaciers break off into the sea, they can create floating islands vastly bigger than any iceberg. In northern Greenland the Petermann glacier has calved some huge ice islands over the past decade. One of the largest, a forty-six-square-mile island that came to be called Petermann Ice Island, broke off in 2010 and began a zigzag journey west and north toward the Canadian Arctic. For a part for the way it was accompanied by the BBC's Helen Czerski, who described Petermann's landscape, in very English terms, as "like a mini version of the South Downs": "gentle mounds were separated by valleys, and these led down to waterfalls of melt water cascading into the ocean."

The most accurate descriptions of natural ice islands date back to 1955, when American scientists lived on another one that snapped off from Greenland. From April all the way to September they mapped every last gulley and crest, along with all its hitchhiking animals and rocks, and their island followed the same zigzag pattern, up and westward. And it had the same fate as Petermann Ice Island, eventually breaking up, many of the lumps getting grounded and stuck fast in the polar wilderness.

Whereas nature's erratic ice islands bump around and head off into oblivion, manufactured ice islands are designed to stay in one place. Engineers were first attracted to ice as a building material because it floats. The earliest plans were some of the most ambitious. In the 1930s a German designer called Dr. A. Gerke came up with a plan for floating ice airports based in the mid-Atlantic. In October 1932, *Modern Mechanix* magazine reported that Dr. Gerke had "erected an ice island in Lake Zurich by artificial means, which endured six days after the refrigerating machinery was switched off." Although Gerke's idea wasn't taken up in Germany, a few years later it was developed by the British scientist Geoffrey Pyke. In the early 1940s he experimented with building an aircraft carrier out of a mixture of ice and wood pulp, which he called pykrete. A prototype was constructed in Patricia Lake in Alberta, Canada, but Pyke's scheme also stalled at an early stage. Unfortunately, although both Gerke's and Pyke's plans have a geeky charm, they were wildly impractical because the seas of the temperate zone are just too warm for ice islands.

Ice has been used for many years as a material for temporary buildings. From igloos to grand houses, such as the ice palace built in St. Petersburg in 1740, it has long been valued as a lightweight, strong, and—if you are in the right place—cheap material. But it was only when detailed plans for the first spray ice island were patented in the United States in the early 1970s that islands made of ice started to attract serious money. Nipterk P-32 was preceded in the 1980s by Mars Ice Island, Angasak Ice Island, and Karluk Ice Island. While they established that spray ice technology worked and that it could provide a stable platform, Nipterk took the technology into uncharted territory.

With a total volume of 860,000 cubic meters, it was nearly twice as big as the largest of the older islands. Moreover, it was situated well out to sea, beyond the protection of barrier islands, in a region where the sea ice can move ten meters in a day. To build it, engineers had to wait until winter, when ice roads could be laid up to the site that were firm enough to allow four spray pumps to be put in position. Spraying started on November 28, 1989, and the temperature hovered around minus-20 Celsius, cold enough for hard ice to be formed in the air and packed into place with bulldozers. For a couple of weeks it got so cold that the ice roads cracked open. There were other unknowns: Would an island this size stay put? Would it split apart if it got too cold? Would it be smashed to bits by the surrounding sea ice? Given these risks, the stability of the island was monitored day and night, enabling the engineers to see how much Nipterk was being compacted and shunted, the biggest lateral movement being measured in millimeters. Nipterk, completed in fifty-three days, was a stunning success, and it was soon able to support a rig as well as service and housing structures.

A new island-building technology had been invented, developed, and shown to work over a period of a decade and a half. And this was not the only type of island that the oil companies created. In the Beaufort Sea, where much of the most innovative activity has happened, there are a dozen or so sacrificial beach islands, which consist of scooped-up beach debris. There are even more gravel islands and various caisson islands, which involve concrete or steel structures being sunk into the sea. There are even rubble spray islands, which are hybrid gravel-and-spray-ice constructions. Nowhere else on earth can one find so many new island-building technologies or see them advance so rapidly. Many of these islands are far bigger than Nipterk. Endicott Island, which lies in Alaskan waters, covers forty-five acres and is made up of two gravel islands linked to the shore by a causeway. But dumping heavy stuff into the sea is expensive. Spray ice islands are about half the price of gravel islands, and they are also cheap to decommission. For a while they looked like the future. Today, however, they look more like victims of their own success. As sea temperatures rise, the Arctic is losing its ice cover and the Beaufort Sea is getting deeper and stormier. Even in October and November large areas of the southern part of the Beaufort Sea are ice-free. The temperate zones where Dr. A. Gerke and Geoffrey Pyke imagined their ice islands are creeping north. As the world heats up, the ice island is headed back into the realm of speculation.

The Floating Maldives

It used to be thought that "living on the water is just for the poor," the young Dutch architect Koen Olthuis said at a public lecture at the University of Warwick in 2012. But as populations expand and sea levels rise, a new attitude is needed. Olthuis works for Dutch Docklands. Founded in 2005, the company has established itself as the market leader in floating island technology. Its projects range from a floating ice hotel in Norway to the Floating Proverb, a planned group of eighty-nine floating islands around Palm Jebel Ali in Dubai that will spell out a poem written by the country's monarch, Sheikh Mohammed bin Rashid Al Maktoum: "Take wisdom from the wise . . . It takes a man of vision to write on water."

But it is in the Republic of the Maldives that Dutch Docklands is making its name. There is already a healthy trade in the floating houses dotted along the Ocean Flower, a flower-shaped raft of planned luxury villas that will be twenty minutes by boat from the capital, Malé. The poor don't come into it: prices for the smallest unit start at $950,000. "The Ocean Flower is an excellent opportunity to gain an outstanding return on investment," runs the Dutch Docklands pitch, especially once you begin "renting out your property through a five star hotel operator."

Ocean Flower is just the start of it. The Maldivian government has signed a contract with Dutch Docklands, leasing to the company four other lagoons around Malé Atoll for a period of fifty years. Using massive rafts pumped with foam and concrete, Dutch Docklands is planning to build an assortment of floating shapes and sizes, including a floating golf course that spans a mini-archipelago, and the Greenstar, a star-shaped and tiered green island that houses a luxury hotel in its upper layers. Apparently the star shape has significance, since, in the words of Dutch Docklands, it "symbolizes Maldivians' innovative route to conquer climate change." They add that it "will become the number 1 location for conventions about climate change, water management and sustainability. A unique Floating Restaurant Island will be built next to it."

Now the super-rich can look forward to flying between their increasingly valuable floating properties while simultaneously saving the planet from environmental catastrophe. It's a trend summarized by a recent *Time* headline: "Floating Technology Will Turn Rising Seas into Prime Real Estate." But according to Olthuis, the rich are just a convenient resource. A young man with long black curls and a boyishly earnest manner, who has already convinced a lot of people that the role of the rich is to "pay for the innovation for the poor," the architect wants nothing less than a complete reorientation of our attitudes toward building space that happens to be covered with water. He used his University of Warwick lecture to argue that "water is a workable building layer" and that "if you turn water into space, which is a dramatic change of mindset, there's a whole new world of possibilities." For Olthuis it is all about getting beyond the static city. With movable floating platforms the "hydrocities" of the future can be as "flexible as a shuffle puzzle." Perhaps an even better analogy is Olthuis's idea that water platforms should be seen as "city apps," each doing a specific job—some for leisure or for sport, some for eating out or for providing trees and wildlife. Each could be called up and deployed when needed, creating a versatile and expandable landscape.

It is a vision that would make not just existing maps but the very idea of maps obsolete. If a city is composed of blocks that can be dragged somewhere else at any time, then old-style visual devices like maps, which are designed as snapshots of a static city, will have to be replaced. Perhaps the new forms of geographical representation that we will need will be more akin to airport gate numbers, digits and time frames that are quickly called up and just as soon forgotten.

Floating villages seem to be coming of age. The first one to be built was in Okinawa, Japan, in 1975, and called Aquapolis. Declared "A Small First Step Toward a Future of Limitless Possibilities," Aquapolis was a thirty-two-meter-high world's fair exhibit that was also designed to function as a self-contained marine community. Twenty years later it was towed to Shanghai and sold for scrap. But Japan continued to build floating structures, and the longest so far has been Mega-Float, a 1,000-meter airport runway that sat in Tokyo

Bay. In 2011 another Mega-Float was used to store contaminated water from the Fukushima Daiichi nuclear power plant.
The largest set of floating buildings ever completed is anchored in the Han River in Seoul, South Korea. It's a giant
conference and events center made up of three connected islands. At the other end of the spectrum there is Joyxee Island,
a tiny private island built by British expatriate Richart "Rishi"
Sowa off the coast of Cancún. It is made from about 100,000
recycled plastic bottles and supports a small house.

For the people of the Maldives one would think there were
practical and immediate reasons for building floating structures. No part of the 1,200 islands that make up the Maldives
is more than six feet above sea level, meaning that the whole
place could soon be under water. It's odd, then, that all the
Dutch Docklands projects are about bringing new people in
and not finding homes for those living there now. The government of the Maldives seems to see the rich in about the same
way that Koen Olthuis does. But the government has a very
different take on how the money they will generate should
be used. A few years ago the prime minister revealed plans
to buy up land in India, Sri Lanka, and Australia. The idea
was that a "sovereign wealth fund" could be built from tourism revenues and a "New Maldives" established overseas. The
tourists would stay and the Maldivians would leave and collect the rent. "Our actions will be a template, an action kit for
other nations across the world," the prime minister boasted.
Yet the prospect of repopulating one's country with floating
pseudo-environmentalist millionaires, while the indigenous
people are transformed from a sovereign body into an ethnic
minority living thousands of miles from home, did not have
wide support. A more popular solution is for the construction

of new islands for the locals using sand dredged from lagoons. The first phase of this process has already been completed. A large new island, called Hulhumalé, has been built using tried and tested methods, and it is one of the highest, and hence safest, places in the Maldives. When the island is fully developed, the government hopes that it will be able to house a third of the Maldives' population. Yet the rise of Hulhumalé casts a further shadow over the pretensions of Dutch Docklands's floating islands to be anything other than playgrounds for the super-rich.

The people of the Maldives seem to prefer solid ground to floating islands. The floating villages on Lake Titicaca in South America and those found across East Asia may be tourist attractions today, but traditionally they have been home to people with few other options. This was certainly the case for the Yau Ma Tei shelter in Hong Kong, a floating boat community of refugees. Bigger examples include Sandu'ao in China, which has its own floating postal service, convenience store, police station, and restaurants, and Kampong Phluk, a collection of attached stilt villages in Cambodia. Both places might be taken to resemble Olthuis's "app city," since their constituent parts are often added to or separated. Yet even in the relatively shielded environment of lakes, floating structures are vulnerable: houses flood, materials decay or wash away, and there is a constant struggle to keep food dry as well as to obtain clean, fresh water.

One of the places where these dilemmas are being seriously confronted is in Singapore. Singapore is a small city-state with a burgeoning population where leading voices are today proposing floating platforms as a cheap way of creating more landmass (cheap, that is, when compared to land reclamation, a laborious process that, since 1959, has increased the size of Singapore by 23 percent). However, there is little dreamy utopianism about these plans, for, as the leading authority on the new platforms, Professor Wang Chien Ming, argues, no one platform could last more than a century. "You may think that a good structure must last 100 years," he says, adding, "Nobody would want to live there after 100 years." It's a sobering thought for those who imagine that rising seas and floating technology are the sure-fire ingredients for prime real estate. How long can floating houses be sold as a luxury option? It turns out that there are good reasons why living on the water was once confined to the poor. In many parts of the world, the unaffordable dream houses of the future may well be the ones built on solid, high ground that is a long way from the sea.

The World

At what point does a ship get so big that it is no longer merely a means of transport but a real place? *The World* is a huge private cruiser that has been circling the world since 2002 and has become a home away from home for its residents. It's also a floating gated community, an enclave of affluence. Perhaps it's also a plush lifeboat, full of refugees from the rest of us.

At home I have a few minutes of Super 8 film of myself, at age three, on the upper deck of the SS *Chusan,* dressed in a crisp white shirt, tie, and smart checked trousers. It was 1967, and my family was on its way to Canada via the Panama Canal. The *Chusan* was scrapped in 1973, one of the last of the old-style ocean liners, and for a decade or so it looked as if ship travel was going to be fully supplanted by air travel. But it seems that people love boats too much for that to ever happen. Over the last thirty years cruise holidays have taken off, on ever more grandiose vessels, and currently the biggest one can carry more than six thousand passengers. *The World*, launched in 2002, isn't on that scale, but it has unique aspirations. Its "130 families" own their apartments and together own the ship. Short holidays on board have been possible, but the real selling point is that *The World* allows you to "travel the world without leaving home." The idea that *The World* is owned and controlled by the residents is reinforced by the fact that the ship's itinerary is determined "collectively."

The World is very expensive. Apartments range in price rom $2 million to $7 million. On top of that, owners pay an annual maintenance fee, which is 6 percent of what their apartment cost, as well as on-board expenses. The owners' identities are well protected, though we do know that *The World* is where the mining billionaire Gina Rinehart, whose father prospected for and discovered asbestos at Wittenoom, spends some of her "downtime."

The World tours the planet in private and isolated splendor. It is both the ultimate adventure and the ultimate secure community, catering to the two seemingly incompatible desires of the ultra-wealthy: to live in pampered seclusion and to drink deep of the very best the earth has to offer. *The World*'s brochure promises a "life of spontaneity. An enchanting and intriguing life. A passionate and adventurous life." It's a distillation and fulfillment of the art of being rich.

It seems to work. This is "the first time I have seen privileged people visibly happy," noted one French journalist after a short stay on board. Other reactions to the same spectacle have been, predictably, more critical. *The World* fits nicely into Robert Frank's notion of "Richistan," a label that tut-tuts disapprovingly at the high jinks of the super-rich and also hints at something more important, their increasing geographical segregation. Sociologists Rowland Atkinson and Sarah Blandy describe *The World* as "secessionary affluence." They see it within a spectrum of economic enclavement, which ranges from small things, such as a Buick SUV called Enclave, to the growth of private jet use and the "mobile mansions" being built by Boeing. For many the self-isolation of the rich is emblematic of everything that is wrong with our era of public squalor and private wealth. Soon after its launch, *The World* was being criticized for its exclusivity. One undercover British reporter claimed that a "deep gloom pervades the ship. The atmosphere is funereal; you'll find more ambience in an out-of-season seaside resort." Residents complained bitterly to her of the holidaying riffraff on board: "How would you feel if you spent millions for an apartment and people who had spent a few hundred pounds had access to the same facilities as you?" Many of these complaints are from the short period before the residents rebelled and took over full ownership in 2003.

The rise of Richistan may be a social disaster in the making, but *The World* should be acknowledged as a pioneer. One of its more original appraisals came from visitors from the Seasteading Institute, a San Francisco–based nonprofit group that advocates floating cities and also enthuses about Sealand (see page 188). They look at *The World* as a forerunner and are keen to learn from its mistakes. Their summary is terse: "beautiful, inspiring, elegant, and wasteful." What really struck the Seasteaders was how much of the public space on board was not being put to good use: "The upper deck tennis court had puddles and piles of ship's supplies next to it. There are 5 or 6 restaurants, but only 2 are open at a time, because demand was just not that high." It seems that this particular "floating city" would have worked better on a more modest scale. They go on to argue that although the ship's low occupancy rate doesn't bother the residents, it does suggest that future projects need to be wary about succumbing to the cruise holiday cliché that bigger is better.

To return to the tennis court, *The World* is the only ship with a full-size court, which is one of its many unique selling points. But it's a feature that highlights the futility of trying to turn ships into real places. After all, a full-size tennis court is hardly a big deal on dry land. It's only when you find yourself somewhere inherently cramped that it becomes a luxury, and that's true of many of the features on the boat. The restaurants, the theater, and the spa become special and glamorous simply by being stuck out at sea, but the games of tennis aren't likely to be any better or the meals any tastier. To take what would be fun for a week and turn it into a lifestyle seems like an error of judgment. The only other people who spend long periods of time in transit are those who have no choice: refugees, salesmen, sailors. Confined and often uncomfortable living conditions are an inevitable consequence of living on the go. *The World* is an aspirational home, but it can never be a real one.

It seems unlikely that what remains essentially a mode of transport can ever develop a sense of community. It might be objected that the ultramobile rich aren't interested in community, as shown by their propensity to erect large walls between them and us and drive in fortified 4 x 4s. But they should be, because the alternative is a footloose existence swept clean of authentic histories and relationships. The kick to be got from living on a big boat with swinging chandeliers and tennis courts is obvious. But such vessels can only ever offer transitory and labored simulacra of what ordinary places achieve effortlessly.

EPHEMERAL PLACES

Hog's Back Lay-By

51° 13' 33" N, 0° 40' 25" W

Many of the most unkempt and unruly places spawned by the geographical imagination are also the most transitory. They are conjured out of very little and sit lightly on the earth, often unnoticed by passersby. These ephemeral places may take the form of improvised townships shaped by or for refugees or for the increasing number of highly mobile workers that modern economies require, but they can also be happy places of temporary escape: a festival city, a child's secret playing spot, or the Hog's Back lay-by.

The Hog's Back is a pleasant grassy hill near Puttenham, a village in Surrey's commuter belt. In a letter to her sister from May 20, 1813, Jane Austen writes fondly of the surrounding landscape seen in fine weather: "I never saw the country from the Hog's Back so advantageously."

Today the Hog's Back lay-by and its surrounding fields and woodland are renowned for dogging. According to the website Let's Go Dogging, "Dogging is a global phenomenon that often involves outdoor sex in car-parks, wooded areas and the like." Let's Go Dogging has conferred on the Hog's Back the title of "second-most popular dogging site in Europe." A sister website, Swinging Heaven, is also a big fan, claiming that "this place has been used for over 50 years as a sexsite." Swinging Heaven lists it along with sixty-one other dogging spots in Surrey, which include some secluded areas in Great Windsor Park, an extensive acreage owned by Queen Elizabeth II.

The word "dogging" derives from a lie that became a euphemism. The idea is that men and women would pretend to be "just taking the dog for a walk" as a cover for nipping out for *en plein air* or in-car sexual encounters with strangers. More interesting is understanding why such a place, with its convenient parking lot but also with its fresh breeze, soft grasses, and mossy wood, sets the blood pumping. Turning ordinary landscapes into what the British police designate as public sex environments (PSEs) implies a hidden and deeply kinky side to geography.

It was David Bell who introduced me to the relationship between sex and geography. David is a tall, elegant professor at the University of Leeds with a waspish sense of humor. In 1994 he delivered a paper to the Association of American Geographers called "Fucking Geography." In 2009 he reprised his theme with an article for one of the learned journals: "Fucking Geography, Again." Professor Bell's insistence that geography should be fucked, not once but until its eyes bulge, has been met with much nervous coughing and feigned indifference. His argument turns on the need for geography to take sexual desire seriously, as something that both shapes people's spatial behavior and might shake up the discipline's conservative bent. To this end he has studied the geography of dogging. It turns out, however, that his real interest is in appropriating dogging into a vocabulary of resistance and subversion. Other academics are pursuing the same angle. For Dr. Fiona Measham of Lancaster University, dogging is "evidence of a continued desire by some to take risks and to resist the regulation, containment and commodification of physical pleasure."

But the idea that outdoor sex is an act of resistance seems contrived. It tells us more about the political desperation of social scientists than it does about dogging in places like the Hog's Back. A more promising approach is pursued by Rowan Pelling, onetime editor of *Erotic Review*. What is it about those "tempting beds of moss"? she asks, before concluding that the "countryside acts as a powerful aphrodisiac." It's an idea with an impressive cultural pedigree. One of the oldest forms of literature is the "meadows of love" poetry found in ancient Greek pastoral lyric. It's a poetic tradition that doesn't just throw in urgent streams, crushed flowers, and sweat-flecked horses as a backdrop. Nature isn't only a picture frame for sex—it *is* sex. The controversial theologian David McLain Carr has recently taken the argument a little further by arguing that Eden, the first garden, was an erotic landscape— paradise as both a tease and a turn-on. But the sensuality of the fields and woods predates religion; it's atavistic, exciting, and excessive. It calls us back.

A Puttenham resident, Ms. Perkins, wearily describes the scene near her home to a visiting reporter: "There were two blokes sitting side by side, watching a man and a woman having sex. Nearby, there were two men sunbathing together, wearing nothing but tight little white underpants." Ms. Perkins was unimpressed by the response of the local police. When she handed them a pink vibrator she had retrieved from the shrubbery, "They said, 'What should we do with it?' I said, 'Put it in Lost Property.'"

There are hundreds of PSEs throughout the UK, but I doubt this reflects a uniquely British inclination for outdoor sex with strangers. It's more likely to indicate a uniquely British inclination to create bureaucracy. For a place to be awarded the title does not mean public sex is officially tolerated but that "PSE-trained" police know about it and will manage it and discourage its use. Although dissuasion is claimed to be the ultimate goal, the police have a duel responsibility toward PSEs. They are aware of the need not to drive those who frequent the Hog's Back, especially gay men, into more secluded and potentially dangerous meeting spots. So they can send rather mixed signals. Between May and July 2010 the Surrey police spent £124.93 on tea, coffee, and biscuits as part of a liaison exercise with Hog's Back doggers. However, recent efforts to make the area less attractive for sex visitors—which center on the Puttenham Parish Council's attempts to turn the lay-by into a wildlife haven—may have had some success. Councilor Richard Griggs claims there are now only "a couple of pockets of activity in the shrubbery."

I wanted to put such claims to the test, so I went to have a look myself in April 2013. The bright yellow gorse was out, the day was bright, and like Jane Austen, I thought I'd never seen the Hog's Back "so advantageously." It didn't take me long to find clumps of condoms in the woodland and an alarming number of surgical gloves. Moreover, individual police officers appear to feel that, like it or not, this particular PSE is here to stay. One dogging chatroom has posted a number of recent comments on how amiable the local police appear to be:

> A friend and I whilst returning from a failed meet had reached that point where we could wait no longer and had pulled over fairly near the Hogs Back layby . . . it was dark and late and whilst cars whizzed past I was busy er . . . hmmm . . . anyway the next thing I notice a Policeman was shining his torch in through the window . . . we stopped and wound down the window a little red faced and he just asked was I ok and reminded us we shouldn't be doing it there but he actually DIRECTED us to the Hogs Back a mile or so up the road if we needed to continue!

My own perverse persistence is to keep trying to find the link between sexual desire and place — more precisely to understand what it is that excites us about having sex outdoors. The doggers at the Hog's Back don't make it easy to work this out, since they appear to think of it as little more than an outdoor orgasmatron. But their thin and sometimes brutal narratives aren't going to dissuade me, for they are clearly taking sensual delight in the open air and bucolic woodland. Geographers still seem unwilling to acknowledge this powerful form of seduction. In a discussion paper pondering his impact on the discipline, Professor Bell concluded with a resigned sigh, "The question, therefore, remains: was geography fucked? I'm not sure it was." But the Hog's Back turns the naughty rhetoric around. It's not what we do to place, it's what place does to us.

LAX Parking Lot

33° 56' 14" N, 118° 22' 15" W

Once upon a time, transportation and destination were very different things, the former merely being a way to get to the latter. But we fell in love with mobility, and today it is often not clear if the traffic is serving the place or the other way around. J. G. Ballard's 1997 prediction that "the airport will be the true city of the 21st century" is already coming true. It is increasingly accurate to talk of transportation networks being fed by places, the classic instance being roadside sprawl, those non-place urban realms that provide a complete support system for, yet are subsidiary to, the demands of travel.

As we forget what we once intuitively understood, the point of real places, it becomes ever easier to be convinced that mobility—ceaseless, on-the-go motion—has intrinsic value: that going *to* places is more important than being *in* places. You could object that this geographical version of the "man versus machine" argument is a nostalgic reaction from someone who can't cope with the fast pace of modern life. Maybe so, but the counterargument is itself now of considerable age, a throwback to modernity's glory days. It hasn't caught up with the fact that where once such worries were based on speculation and dystopian fantasy, they are now evidenced by looking out of the window. What you see is that places are atrophying as routes and roads swell. Parking Lot E at Los Angeles International Airport, commonly referred to by its acronym LAX, takes our window view just a little bit further. A few thousand feet from the end of runway 25L resides a new kind of community.

Most of those who live in the RVs that occupy the eastern end of Lot E aren't there permanently or even all week. They form a commuting settlement made up of pilots, mechanics, and flight attendants, many of whom hitch lifts from airlines to get to work, often bunking down again for some extra sleep at their destination. Airline safety regulations insist that crew turn up well rested, but for employees that is not as easy as it sounds. Most airlines have moved to a business model that sends their staff all over the country and farther afield. The old system of offering people transfers, and paying for them and their families to relocate, to settle down in a new place, has been replaced by something much less expensive and much lonelier.

The lot, which is limited to one hundred vehicles, has been formally recognized by the airport authority since 2005. Residents pay $120 a month to park their RVs there, with an additional $30 fee for their cars. It's cheap if not very cheerful. And the few newly planted rose bushes can't the hide the fact that this is a place of last resort for an industry that has squeezed wages and working conditions. One resident interviewed for National Public Radio reported that he hadn't had a pay raise in twenty years: "It's always 'you need to take a pay cut,' 'you need to take a pay cut,' 'you need to take a pay cut.'" A neighbor makes the same point: "It's been a devastated industry. Things are not what we thought they were going to be."

The NPR story annoyed residents because it named the interviewees. Most would prefer to keep their anonymity,

since Lot E is not a place they are proud of. One pilot explained to a visiting reporter, "I never thought I would be here,
but pay cuts force us to be frugal." In another interview, a
neighbor vents understandable bitterness: "Pretty glamorous,
isn't it? I, for one, never thought I'd end up at a parking lot at
LAX." Many residents cling to the idea that they are not really living at Lot E but merely using it like a glorified locker
room. One pilot with a house in Texas says it is just "a place
to come and get ready for work." Yet like so many others, he is
geographically trapped, a long way from work and a long way
from home. What might at first seem like something temporary and convenient can easily turn into something semi- or
completely permanent.

Considering they are employees who provide a vital service
for US airlines, the Lot E villagers are treated pretty shabbily.
The airport does not provide electricity, propane, or water.
The residents have to be canny in order to acquire basic services. They rely on solar panels, small generators, and showers
taken at the local gym. It is a Spartan existence and its challenges are added to by the roar and lights from the planes that
are landing almost on top of them. Some airline employees
take a stoic delight in all the noise. "I love to see what's coming in," one worker told NPR. "It doesn't worry me—I love it.
I get a thrill." Noting that flights start at exactly 6:30 a.m., his
neighbor dryly offered another plus: "You don't need an alarm
clock." But no amount of chipper dedication can disguise that
the noise is almost unendurable. Other residents have taken
to covering their windows with foil and paper to muffle the
sound, or playing recorded white noise in their RVs, a static
rumble that takes the edge off the shriek of the planes.

Not everyone dislikes the idea of airport living, and some

are even willing it on. Professor John Kasarda of the University of North Carolina travels the globe extolling the pleasures as well as the inevitability of the "aerotropolis." For him LAX is the center of town: Kasarda regards the essence of a modern place as the availability of a flight to somewhere else. There are good reasons to resist this twenty-first-century re-kindling of the Corbusian dreamscape of speeding machines swarming through geometric space. The motorized land-scapes created in the twentieth century taught us that this vision doesn't meet human needs, nor does it create real places, which comes to the same thing. We want places that are worth journeying to rather than non-places that are byproducts of a ceaseless need to keep moving. In the face of the flow of modern history, real places—places with diverse and complex human histories, places where people come first—have taken on an oppositional character. They are engaged in, or poised for conflict with, the engorged but ever greedy traffic. It's a stark choice. The need to tilt the balance of power back away from travel and toward place is plain.

Yet a sense of inevitability, of submission to the iron will of something bigger than any of us, seems to have been injected into the cultural bloodstream. How else to explain our readiness to believe the endlessly recycled story that "the sector" is in "crisis," and not just the airline sector but any and all sectors of business. That unless we give way, become more flexible, go to contract work, and move into rental units many miles from home, we soon won't have any airplanes or cars or jobs. Amid the surrounding din one of the Lot E residents explained on NPR: "It's an industry in the throes of stagnation and maybe the early throes of death. Maybe in 10 years, the airlines won't even be here anymore. It's that bad." It's true;

things are bad. But it's also true that we have gotten so used to messages about adapting to "crisis" that the inhuman demand that we live as rootless nomads has become difficult to challenge. The normalization of "crisis" has created the conditions for people to give up on things that matter to them, like real relationships and places worth flying to. The non-places created by this restless movement feed the traffic and keep the wheels turning. Yet they are so subsidiary to mobility that they also resemble parasitic growths, latched onto an indifferent host.

Nowhere

41° 41' 49" N, 0° 10' 12" W

The Nowhere festival is held every July on a dusty plain in Aragón, in the north of Spain. It's a temporary utopia; although you need to buy a ticket to get in, once inside, the Nowhere website announces, "you cannot buy or sell anything, except for ice, for just a few hours a day, so you don't suffer food poisoning or warm beer." The result is a creative economy of "radical self-expression" and "self-reliance." It's organized around a cluster of camps where everything and anything can happen: pageants, Japanese tea ceremonies, erotic life drawing, circuses, as well as lots of music and art. Participants are encouraged to make full use of the site's Costume Camp: in the words of the site blog, "think of it as a giant dressing up box and release that inner child!"

The idea of festivals as places apart is laden with libertarian messaging and rooted in 1960s counterculture. Over the past two decades this sensibility has come in from the margins and blossomed into a mass phenomenon, and the result has been to open up a new chapter in the idea of place. To see whole communities suddenly take shape out of nothing, in the middle of nowhere, is fascinating for a place-loving species. The thrill is multiplied when that new place is bound together by a shared attachment to novelty and autonomy.

Since places like Nowhere, and its older and much bigger Californian cousin Burning Man, were founded, even remoter and odder events have come along, such as the Traena Festival, which takes place on a Norwegian island inside the Arctic Circle. In fact, some musicians and fans find this island too accessible, and so an offshoot event has formed. This clique all head off to Sanna, a bleak rock slapped by the wind, in order to play and listen to music in a sea cave.

Festivalgoers have developed a taste for geographical extremism, and distant and spectacular spots that are challenging and hard to get to are much sought after. Because of threats by Islamist militants, the event once promoted as the world's most remote, the Festival au Désert, usually staged near Timbuktu in Mali, went into exile in 2013 and took place in Burkina-Faso instead. But it plans to be back, and an audience will come. Distance is a way of sifting the crowd, making it unlikely you'll have to share acts of self-expression or a food tent with people who are dissimilar to yourself. But that's the sort of sly dig you might expect from a self-confessed festival avoider, especially one whose sarcasm is whetted by envy. The creation of Nowhere is actually a charming mixture of the imaginative and the prosaic. A lot of thought and care are clearly given to how to make this small patch of

desert into a happy community. Place-making demands attention to the everyday practicalities of living as well as the big
picture. Ironically, it is these prosaic details that churn and
excite our imagination: by sorting them out we prove that new
places are within reach, that we are capable of conjuring up a
new world in an empty landscape.

The first thing that gets built at Nowhere is Werkhaus, an
operational base of toilets, kitchen, and shade. Then attention
turns to the festival's hub, the Middle of Nowhere. The need
for a central meeting area was understood from the first Nowhere, in 2004. It was built of piping and parachute material
and sited at one side of the camp. That last detail may seem
insignificant, but it turns out that it didn't quite work: places
need centers that are located at the heart of things. So in 2009
the Middle of Nowhere was moved to the physical center of
the festival. The energy and mayhem of Nowhere thrives on
careful decision-making and clear lines of communication
and authority. Without the individuals who take on the responsibility of being a "Nowhere Lead," the place would become disordered, people would drift away, and the fun would
fizzle out. Nowhere deconstructs and overturns the conventions of mainstream places yet relies on a strict division of
labor and responsibility in order to fashion a workable alternative. Each Lead coordinates a different function, working
with other volunteers leading up to, during, and after the festival. These functions are reassuringly mundane: toilets, tick-
ets, power grid, recycling, safety, medical team, and the like.
There is also a communications infrastructure, which centers
on the site's post office and the daily newspaper, *Nowhere Tri-
bune*.

Much of the pleasure of creating places is in getting such

details right. As "Europe's answer to Burning Man," Nowhere has a model it follows. Yet the paradox about Nowhere is that, although "radical self-expression" is the end product, what fuels all the hard work and unpaid hours is not just an ambition to be original or express oneself but the innate human love of place. This is why being able to shape a working, living place from scratch, and then see it filled with excited participants, enthuses enough people to make it happen every year. The thrill of creating Nowhere seems to be heightened by the "leave no trace" policy that it took from Burning Man. At first glance this sounds odd, since we are used to the art of place-making being identified with permanent structures. For many generations we have been establishing new places by sinking foundations, grasping for immortality with the weight and endurance of our grand designs. But in postindustrial societies that no longer have faith in either architectural monumentalism or their own future, this is no longer a convincing conceit. And Nowhere seems to turn the edifice complex on its head with little effort. It's hard not to be impressed by the way the organizers and participants can craft this place then fold it away, as if by magic, leaving a pristine and empty landscape. It seems that places don't need to be long-lasting impositions in order to be important or substantial. Indeed, the ongoing growth of festivals suggests the opposite: that there is something about a place that can quickly disappear that adds to its aura.

This eco-generation's wisdom on the topic is not always appreciated. A lot of the alternative festival scene in contemporary Spain is supported by British expatriates who felt driven out of their own country by antifestival legislation enacted in the 1990s. The so-called New Age Travellers, who used to wander from site to site in the UK, either ended up as squatters or went abroad, with many heading for the warmth and empty spaces of southern Europe. Since then festivals have multiplied, and far from being despised, people who can sort out and kick-start successful events are in great demand. Even militantly anticommercial festivals like Nowhere, with its explicit rejection of the cash economy, have settled down into valued cultural assets. We have come to appreciate these small utopias, not necessarily for the music or the face painting, but because they remind us of something that we never meant to forget, that making places is serious fun.

Stacey's Lane

51° 41' 48" N, 0° 06' 57" E

Children create their own places in between and around the adult world. As children, my brother and I made our dens in Stacey's lane, a blind alley named after the family who lived at the far end. Behind their house was a set of rough fields broken up by patches of beech and birch trees where Paul and I would also build camps. To look at it today, you wouldn't think the lane would be worth hanging around in, since it is only a hundred meters long and lined with thin, dirty-looking trees and straggly bushes on either side. These side strips are only a meter or two deep, often less, and behind them there is an assortment of tall fences and messes of wire. It's a gloomy place whatever the weather, always cold and muddy. But when we were little this is where we created places of adventure and escape. We had four or five hidden spots on the go at any one time, made by breaking back the branches and flattening out just enough room to stand in together. Having a number of them was somehow important, presumably because it meant there was always work to be done and always somewhere else to go when we argued. It meant we could separate and come together, bringing information or biscuits from home, which was just around the corner on the main road.

Given half a chance, children create their own nooks in the leftover places of the adult map. We weren't interested in the fields beyond because what we wanted were hidden places that would be bypassed by the adult world. Years later I watched my own two children doing exactly the same thing. The rhododendron bushes in my local park provide the most popular sites. If you peer into their tangled gloom, you'd see grubby girls and boys both. There is usually at least one exasperated parent urgently circling them, for the bushes are not just for den-making but for evading adults.

In *The Child in the City* Colin Ward argues, "Behind all our purposive activities, our domestic world, is this ideal landscape we acquired in childhood." He goes on to describe these lost places as allusive yet persistent: "It sifts through our selective and self-censored memory as a myth and idyll of the way things ought to be, the paradise to be regained." Ward's argument suggests that children's den-making isn't just echoed in later life but that it is constantly being sought after. The secret places may be long gone and rarely recalled, but they offered something so important and so consoling that they remain with us, their simulacra fashioned time and again in our homes or cars, the adult dens that bring us comfort.

Childhood dens are our first places, or at least the first

places we actively shape with our imagination, care for, and understand. The uncomfortable nests that I helped stomp out between branches in Stacey's lane were where I learned that places can be far more interesting than the set of routines and lines of demarcation I was used to being ushered into. I also have a clear memory that they offered more than just a feeling of security or the fun of hiding. I can recall the whispered conversations between Paul and me, conversations that remade the significance of each den over and over again: this is your base; no, it's my main base; no, this is an entrance to those two dens, which are bedrooms. The meaning of each place was entirely at our disposal and was constantly being transformed to fit with our changing fantasies.

Den-making is a particular kind of play, not with dolls or toy guns but with place. It's a form of play that is particularly private and vulnerable. Any adult or teenage presence can destroy it at once: a looming face would reduce Paul's and my dens to a dull clutter of sticks. For adults it's hard to recapture this ephemeral, playful approach to place-making because, as we grow up, we get used again to the idea that the meaning of places is fixed and not ours to command. Many of the other entries in this book—the micro-nations, the remote festivals, the all-male religious communities—strike us as remarkable because they resemble den-making. Yet, when compared to children's dens, they are static places, always frozen in one imaginative moment.

But even as Paul and I were hunkering down in the bushes with our biscuits, there were other adults who were getting worried that fewer children were being allowed similar experiences. In 1960 Paul Goodman had already described the "concealed technology, family mobility, loss of the country, loss of neighborhood tradition" that were, he said, "eating up" play space and taking away the "real environment." Today the idea that children's play is endangered is widely shared. Writing of urban Australia, the educational researchers Karen Malone and Paul Tranter suggest that "many children have lost access to traditional play environments, including streets and wild spaces." They lay the blame at a number of doors: "parental fears about traffic danger, bullying and 'stranger danger,'" as well as "the loss of natural spaces." All these parental worries mean that the streets seem risky places for children, so too the parks. These anxieties are not baseless, but their consequence is that play is increasingly viewed as a time-limited "experience" to be managed by experts. Professionally designed playgrounds and "play facilitators," "play workers," and "play assistants" are colonizing the territory, which is an impossible and a paradoxical task. On the one hand we want to protect children, but on the other we want them to relive our own imagined childhood adventures. Play professionals are asked to both secure children from risk and introduce them to risk by prying them from the grip of screen-based leisure.

Ironically, den-making is something done on a computer for many children today. I know that if my two children saw the unkempt structures Paul and I made, they would be deeply unimpressed. There are numerous websites that allow them to create, in comfort, not just their own rooms or houses but whole private landscapes and kingdoms. This is often done for the benefit of an avatar, but at root it's a form of den-building, imaginative place-making that appears to be hidden from a nosy and bothersome adult world. Yet these virtual warrens lack something: they do not make their users reassess their relationship to real places or grasp their power to shape them. They are, after all, grown-up creations: manicured spaces with very strict rules and a limited number of options. If it is true that we spend adulthood trying to reconstruct the warm, free, and happy fantasy places of childhood, then it will be interesting to find out how a generation brought up on the Sims can nostalgically rework its computer-mediated memories of geographical play.

CONCLUSION: SYMPATHY FOR A PLACE-LOVING SPECIES

I have introduced just a fragment of a world of remarkable places, one that reveals the range and power of the geographical imagination. I believe it also tells us something about our own relationship with the places around us. Unruly places have the power to disrupt our expectations and to reenchant geography. They force us to realize how many basic human motivations—such as the need for freedom, escape, and creativity—are bound up with place. From Sandy Island to Stacey's lane, we have seen how people pour their hopes and fears into place.

The book's journey, from places that are lost and hidden to places that are designed and crafted, has followed the human instinct to shape and create. Yet none of these encounters has offered reassurance or comfort. We have had to confront some of the oddest but also some of the bleakest and most difficult locations on the planet and learn that their stories matter to us; that what happens to sinking islands and new deserts or towns in the grip of brutal authority concerns every place.

We have also had to confront the fact that our relationship to place is riddled with paradoxes. Ordinary places are also extraordinary places; the exotic can be around the corner or right under our feet. Another striking paradox is the way borders both trap us and give us our sense of liberty. Some of the places in *Unruly Places*, especially those I have labeled as no man's lands, appear to have escaped the claustrophobic grid of nations and, hence, offer a promise of freedom. Yet in their unpredictability and sometimes cruelty, these places also impress on us why people find borders so necessary. The paradox can be deepened: the reason we keep drawing borders is not a matter of mere utility. Borders can inspire and excite us. Breakaway nations provide the clearest example, but it's also an emotion that can be found from Baarle-Nassau to Mount Athos. Not only is a world without borders never likely to happen — it also wouldn't be much fun.

Another paradox that emerges from the forty-seven disorientating places gathered in this book is humanity's need for both mobility and roots. "Among the great struggles of man," Salman Rushdie tells us in *The Ground Beneath Her Feet*, "there is also this mighty conflict between the fantasy of Home and the fantasy of Away, the dream of roots and the mirage of the journey." A few of the places we have encountered try to ride this divide: residential ships like *The World*, which ceaselessly tour the globe, and ephemeral places like the Nowhere festival, which spring up suddenly then disappear without trace. But this is a dilemma that can never be neatly or completely resolved. The lure of escape and wanderlust is just as deeply implanted as its polar opposite, the desire to anchor oneself in a particular place, to know and care for somewhere that isn't just anywhere.

Because place is integral to human identity, so too are the paradoxes of place. People's most fundamental ideas and attachments don't happen anywhere or nowhere; they are fashioned within and through their relationship to place. This may not be a new argument, but it's one we still find difficult to discuss. Or, more precisely, we find it hard to talk about with ambition, as something with intellectual content and reach. Thankfully there are some signs that we are at the start of a period of reflection on the nature of place. I find it telling that a book called *The Destruction of Memory*, by Robert Bevan, is all about the destruction of place, more specifically the last century's depressingly long roll call of places that were bombed or demolished for the sake of militant ideologies of one sort or another. Such authors as Paul Kingsnorth in Britain, Marc Augé in France, and James Kunstler in the United States, who have put the death of real places and the rise of non-places and "nowheres" onto the cultural agenda, are feeding into the same conversation. Yet while those who care about place have a lot to be troubled about, it would be a shame if this discussion was limited to nostalgic laments. As we have seen, the world is still full of unexpected places that have the power to delight, sometimes appall, but always intrigue. These unruly places provoke us and force us to think about the neglected but fundamental role of place in our lives. They challenge us to see ourselves for what we are: a place-making and place-loving species.

Bibliography

Ackroyd, Peter (2011). *London Under*. London, Chatto and Windus.

Alterazioni Video (2008). *Sicilian Incompletion* (*Abitare* 486, pp. 190–207), available at www.alterazionivideo.com/new_sito_av/projects/incom piuto.php.

Augé, Marc (1995). *Non-Places: Introduction to an Anthropology of Super-modernity*. London, Verso.

Babcock, William (1922). *Legendary Islands of the Atlantic: A Study in Medieval Geography*. New York, American Geographical Society.

Ballard, J. G (1962). *The Drowned World*. New York, Berkley Publishing.

Ballard, J. G. (1974). *Concrete Island*. London, Jonathan Cape.

Bevan, Robert (2006). *The Destruction of Memory: Architecture at War*. London, Reaktion Books.

Boym, Svetlana (2001). *The Future of Nostalgia*. New York, Basic Books.

Brick, Greg (2009). *Subterranean Twin Cities*. Minneapolis, University of Minnesota Press.

Brodsky, Joseph (1987). "A Guide to a Renamed City," in *Less Than One: Selected Essays*. London, Penguin Books.

Casey, Edward (1998). *The Fate of Place: A Philosophical History*. Berkeley, University of California Press.

Choisy, Maryse (1962). *A Month among the Men*. New York, Pyramid Books.

Egremont, Max (2011). *Forgotten Land: Journeys among the Ghosts of East Prussia*. London, Picador.

Frank, Robert (2007). *Richistan: A Journey Through the 21st-Century Wealth Boom and the Lives of the New Rich*. London, Piatkus.

Guevara, Ernesto (1961). *Guerrilla Warfare*. New York, MR Press.

Hills, Ben (1989). *Blue Murder: Two Thousand Doomed to Die—The Shocking Truth about Wittenoom's Deadly Dust*. South Melbourne, Australia, Sun Books.

Hohn, Donovan (2012). *Moby-Duck: The True Story of 28,800 Bath Toys Lost at Sea*. London, Union Books.

Kasarda, John D., and Greg Lindsay (2011). *Aerotropolis: The Way We'll Live Next*. London, Penguin Books.

Kingsnorth, Paul (2008). *Real England: The Battle Against the Bland*. London, Portobello Books.

Kunstler, James (1993). *The Geography of Nowhere: The Rise and Decline of America's Man-made Landscape*. New York, Simon and Schuster.

Lee, Pamela (2000). *Object to Be Destroyed: The Work of Gordon Matta-Clark*. Cambridge, MIT Press.

Moorcock, Michael (1988). *Mother London*. London, Martin Secker and Warburg.

Miéville, China (2009). *The City and the City*. London, Macmillan.

Mycio, Mary (2005). *Wormwood Forest: A Natural History of Chernobyl*. Washington, D.C., Joseph Henry Press.

Nunn, Patrick (2008). *Vanished Islands and Hidden Continents of the Pacific*. Honolulu, University of Hawaii Press.

Stommel, Henry (1984). *Lost Islands: The Story of Islands That Have Vanished from Nautical Charts*. Vancouver, University of British Columbia Press.

Tuan, Yi-Fu (1974). *Topophilia: A Study of Environmental Perception, Attitudes, and Values*. Englewood Cliffs, N.J., Prentice-Hall.

Ward, Colin (1977). *The Child in the City*. London, Architectural Press.

Acknowledgments

Many thanks to Robin Harvie at Aurum and Courtney Young at Houghton Mifflin Harcourt for their tireless help, encouragement, and many suggestions; James Macdonald Lockhart for his faith and patience; and Rachel, Louis, and Aphra for their numerous ideas and for listening.

Index

Blandy, Sarah, 223
blue-asbestos mining
 Koegas, South Africa, 97
 See also Wittenoom, Western
  Australia
Bogotá Declaration (1976), 143
Boiler Fires (World War II), 13
borders
 attitudes toward, 177–78, 182
 Conference of Berlin and, 72
 meanings/purposes and, 75, 76–77,
  177–78, 251–52
 See also enclaves; land between
  border posts; Nahuaterique
Bountiful
 Anastasia Movement and, 150–52
 description/values, 149, 151–52,
  153–54
 location, 149
 See also utopia/pursuit
Boym, Svetlana, 10–11
Bradbury, Paul, 110
Brick, Greg, 40
Bright Light interrogation/detention
  center
 CIA and, 137, 138
 description/location, 136–39, 140
 detainees treatment, 139
 high-value inmates, 138–39
 as non-place, 137
British Admiralty map, Sandy Island,
  3, 6
Brodsky, Joseph, 8
Brotas Quilombo
 attitude change toward, 161–62
 issues of past/present and, 159–61
 "ranching sprouts" term, 160
 See also quilombos
Brown, Matt, 71–72

Bucharest's Palace of Parliament
  (House of Ceauşescu), 107
Bulgaria's Buzludzha Monument, 107
Burning Man, 239, 241. *See also*
  festivals as places

Cameron, David, 195
Camp Zeist/trial
 bombing of Pan Am flight and, 128
 Netherlands/Scotland and, 127–28,
  129–32
 outcome, 131
 preparations/outcome, 128, 129–31
 views on, 131–32
Cappadocia
 history/Christians, 48–49
 See also underground cities of
  Cappadocia
Carr, David McLain, 231
Casey, Edward, xii, xiii
Castellanos, Salome, 81
Castrillo, Roberto Hidalgo, 78
Catherine II, Empress of Russia, 105
Ceauşescu, Nicolae, 107
"cellar fishing," 146
cemetery living
 description/examples, 55–59
 See also City of the Dead, Cairo;
  North Cemetery, Manila
Centralia, Pennsylvania, 98
Chamumbala, Don Muatxihina, 194
Charles, Prince, 157
Chayanov, Alexander, 150
Chernobyl nuclear power plant. *See*
  Pripyat
Child in the City, The (Ward), 243
children creating places
 decline in, 244–46
 description/importance, 242–44